Meeting the Alien

Andreas Anton · Michael Schetsche

Meeting the Alien

An Introduction to Exosociology

Foreword by John Elliott

 Springer VS

Andreas Anton
Institut für Grenzgebiete der Psychologie
und Psychohygiene
Freiburg, Germany

Michael Schetsche
Institut für Soziologie der
Albert-Ludwigs-Universität
Freiburg, Germany

ISBN 978-3-658-41316-3   ISBN 978-3-658-41317-0   (eBook)
https://doi.org/10.1007/978-3-658-41317-0

This Springer VS imprint is published by the registered company Springer Fachmedien Wiesbaden
GmbH, part of Springer Nature.
The registered company address is: Abraham-Lincoln-Str. 46, 65189 Wiesbaden, Germany

Foreword by John Elliott

Is anybody out there? A question that has resonated across humanity for generations. Recent advances in technology have enabled us to realize that whenever we see a pinpoint of light in the night sky, regardless of which one we choose to focus on, there is a good chance that there will be a planet nearby. But will we ever receive a message from E.T? Will we find evidence of other life in the Universe? Of course, we do not yet know the answer to these questions. We also do not know when this could happen—ancient microbial life may be buried beneath the Mars rover at the very moment you are reading this. What we do know is that we cannot afford to be ill-prepared—scientifically, socially, or politically—for such a profound event that could turn into reality as early as tomorrow.

My own personal involvement with SETI and post-detection research began in the late 1990s, when my contributions focused on developing methods for identifying and deciphering language-like information content in signals. This journey led me to investigate many different languages (communication systems), both human and animal, to ascertain whether there could be any underpinning common structures that might serve as communication "fingerprints" for recognizing a message from afar. During this time, it became increasingly evident that much was needed to organize our post-detection provision, to ensure that we are as ready as we can be for such an event.

In 2022, along with colleagues at the University of St Andrews, I officially launched the SETI Post-Detection Hub as a permanent "home" for coordinating the development of a comprehensive operational framework. This has brought together interested members of the SETI and wider academic communities, as well as policy experts, to work on topics ranging from message decipherment and data analytics to the development of regulatory protocols, space law, and societal impact strategies.

The science of discovery—exploring both our known and unknown unknowns—challenges us to the limit of our current knowledge, perceptions, and technology, as we endeavor to avoid any limiting human conceit: a process that is ever-evolving. This science also has much to contribute in our quest to answer terrestrial questions, while helping us to understand what makes us human.

In this book, through their (exo)sociology lens, Schetsche and Anton expertly take us on a detailed journey that reflects many of the multidisciplinary challenges we are facing, together with the thought processes and rationales behind them. They examine our scientific investigative history, its extent so far and, in some cases, how it has arguably been limited by professional bias and anthropomorphism, as well as the scientific "hot potatoes" of METI and UFOs. The authors also take us on a fascinating journey of possible contact scenarios and the consequences for, and impact on, humanity: scenarios that also explore our philosophical speculations as well as possible alien physiologies (both biological and post-biological).

Scientific enquiry often shows us that the truths we once held as absolutes can be proven wrong, as we develop better ways of observing. Therefore, we need to realize that our journey is only just beginning and that there is still much to discover and reflect upon. This book goes a long way in presenting and exploring the possibilities that lie ahead.

St Andrews, UK   Dr. John Elliott
                                            Coordinator of SETI Post-Detection Hub:
                                                  https://seti.wp.st-andrews.ac.uk/
                                               Chair of UK SETI Research Network
                                                                         (UKSRN)
                                         IAA SETI Permanent Committee Member

Foreword to the German Edition by Dieter B. Herrmann (1939–2021)

In 1978 I read with some fascination the German translation of the book *The Colonization of Space*[1] by the US physicist and visionary Gerard O'Neill (1927–1992). In those years, manned spaceflight made rapid progress, not least fueled by the competition between the two political systems, the USSR and the USA. The Americans had landed on the moon, the Russians had put their first space stations into orbit starting in 1971, and both nations launched the joint *Soyuz-Apollo* venture (1975) as a symbol of hope for peaceful cooperation in the cosmos. The question was obvious: What would happen next? Would there perhaps even be a hint on the horizon of the vision that Tsiolkovsky had once expressed in the words: "Mankind will not remain on Earth forever, but in the pursuit of light and space will first timidly penetrate beyond the boundaries of the atmosphere and then conquer all the space around the sun" (Kosmodemjanski 1979, p. 178).[2]

As thoroughly as O'Neill had examined the problems of the technical feasibility of his vision, he seemed to me to see far-reaching prospects on one crucial point, the justification for which seemed more than questionable. For he wrote: "More important than material concerns, however, is, I believe, the legitimate hope that the opening of a new vast habitat will bring out the best in us, that the new territory awaiting its creation in space will grant us new independence in the search for better forms of government, social structures, and ways of life, and that our children may thereby find a world which, thanks to our efforts, shall offer them richer opportunities during the decades to come" (O'Neill 1978, pp. 250–251). The question then suggested itself to me: Can the colonization

---

[1] German edition: Gerard K. O'Neill. 1978. *Our Future in Space*. Bern, Stuttgart: Hallwag.
[2] Cited in A. A. Kozmodemyansky. 1979. *Konstantin Eduardovich Tsiolkovsky*. Leipzig: Teubner Verlagsgesellschaft.

of space bring about a fundamental change in the moral constitution of man or, conversely, would not this change perhaps have to precede the mammoth project of colonizing space as a condition of its feasibility?

The focus of today's worldwide discussions is a similar question—only from the opposite perspective. Not only scientists, but every thinking person wants to know whether life is a universal manifestation in the cosmos or whether the technical civilization of mankind is a unique exception that could only emerge and develop on planet Earth through the interaction of extremely improbable coincidences. If the former is true, further questions inevitably arise: What could the "others" be like; how would they look, communicate, think, and act; and how would they be socially organized? Should they not—like us—have developed the technical ability to travel into cosmic space beyond their homeland by means of space travel and perhaps even pay us a visit one day? With what intentions would they come then? If O'Neill were right in his "moral vision", one could conclude: A technical civilization capable of interstellar travel is also vastly superior to us in moral terms and comes with peaceful intentions.

But these are all questions that reach far beyond the realm of scientific research. Neither astrophysicists nor biologists nor chemists can make well-founded statements about them. What already sounded vague and perhaps also somewhat naive in O'Neill's work requires a thorough scientific discourse. The great advances in astrophysical research in recent decades have made it increasingly likely that we are not the only intelligent living beings in the universe. Therefore, addressing the related questions, which have so far been dealt with only marginally, has also gained urgency.

And herein lies the merit of the two authors Michael Schetsche and Andreas Anton. They have spent years of scientific work dealing with the sociological problems associated with the question so often asked, "Are we alone in space?" and call their approach to this new scientific discipline "exosociology", knowing full well that sociology originally has nothing to do with extraterrestrials. But without sociological considerations, it will not be possible to answer the questions now before us. However, the authors do not limit themselves to their discipline of sociology in the narrower sense, but at the same time present the astrophysical foundations on which our present knowledge and more or less well-founded assumptions are based. Thus, also the generally interested reader can benefit from this work, which contains a wealth of suggestions and thus should and certainly will result in further considerations and research.

Berlin, Germany   Dieter B. Herrmann
June 2018

Acknowledgements

First of all, we would like to thank the Institute for Frontier Areas of Psychology and Mental Health (IGPP) in Freiburg, Germany for creating the scientific freedom and the financial support that made this book possible. In particular, we would like to thank Kirsten Krebber of the IGPP team for her highly committed editorial supervision of the volume.

In addition, we would like to thank the following people (in alphabetical order) for support: Luana Arena, Michael Bohlander, Fabian Bornemann, Ulrich Dopatka, Wolfgang Eßbach, Keith Farrington, Tobias Daniel Gerritzen, Nadine Heintz, Dieter B. Herrmann, Kerstin Hoffmann, Jan Holtkamp, Cori Antonia Mackrodt, Thorsten Mann, Alton Okinaka, Bernd Pröschold, Renate-Berenike Schmidt, and Martin Werner.

With regard to the English edition, special thanks are due to John Elliott, who did the English editing and wrote a wonderful preface. We look forward to further academic collaboration.

Finally, we would like to thank all colleagues of the *Forschungsnetzwerk extraterrestrische Intelligenz* (Research Network Extraterrestrial Intelligence)—without this special discussion context, the book would certainly never have been written in this form (https://www.eti-research.net/pdf/eti_positionpaper_eng.pdf).

Contents

About the Authors

Andreas Anton studied sociology, history, and cognitive science at the Albert-Ludwigs-Universität Freiburg, Germany. He completed his Doctorate in Sociology in the German Research Foundation (DFG)-funded project "In the Shadow of Scientism. On dealing with heterodox knowledge, experiences and practices in the GDR". Since 2017, he is a Research Associate at the Institute for Frontier Areas of Psychology and Mental Health (IGPP) in Freiburg. He is a founding member of the German ETI Research Network and a member of the SETI Post-Detection Hub at the University of St. Andrews (UK). Main research interests: sociology of knowledge, media and culture, exosociology. Contact: anton@igpp.de.

Michael Schetsche studied political science with a focus on futurology at the Otto-Suhr-Institut of the University of Berlin. He was a student there of Ossip K. Flechtheim, the father of German "futurology". He received his Doctorate and Habilitation in Sociology from the University of Bremen. He worked for 20 years as Department Head and Research Coordinator at the Institute for Frontier Areas of Psychology and Mental Health (IGPP). For many years he has been an extraordinary Professor at the Institute of Sociology at the Albert-Ludwigs-University of Freiburg. His primary areas of work are sociology of knowledge, sociology of culture, cultural anthropology, futurology, and exosociology. He has dealt with the question of the relationship between humans and extraterrestrials in a whole series of essays since the late 1980s. He is a founding member of the German ETI Research Network.

The Aliens at the Gates of Sociology

1

When one of the authors of this volume gave an interview to the German newspaper *Süddeutsche Zeitung* in 2017 on the subject of 'exosociology', the first question asked by journalist Esther Göbel was: "Now let's be honest: What is that supposed to be?" Of course, since there was a preliminary discussion to the interview, this question was of a rhetorical nature. Nevertheless, it sums up the general problem of this book very well: we have taken it upon ourselves to write an introduction to a sub-discipline of sociology that does not even exist in this form. Up to the present day, at any rate. Therefore, we are forced—rather unusually for sociology—to first explain in some detail what this new hyphenated sociology is all about, what its subject matter and goals are. And, as is customary in academia, we begin with a brief conceptual history.

1.1   The Idea of an 'Exosociology'

The first time the term 'exosociology' appears in scientific literature is probably in an anthology published by the Russian radio astronomer Kaplan (1971) in 1969,[1] which is entitled *Extraterrestrial Civilizations: Problems of Interstellar Communication*. The chapters of the volume were penned by astronomers, mathematicians and linguists—a sociologist was not included. Despite this, or perhaps because of it, the editor places the term *exosociology* in a prominent position in his introduction to the volume. With it he names *all* efforts of a search for the signals of extraterrestrial civilizations—which for him includes three central program points: (1) a general theory of the development of civilizations, which

---

[1] The citation is based on the English translation available to us from 1971, which NASA had prepared.

A. Anton and M. Schetsche, *Meeting the Alien*, https://doi.org/10.1007/978-3-658-41317-0_1

1

allows conclusions beyond human societies; (2) the development of strategies of the search for extraterrestrial civilizations; (3) linguistic problems of decoding extraterrestrial messages. In a nutshell, everything that is called 'SETI programming' in the US research tradition (we will return to this later) is here subsumed under the term exosociology. Kaplan (1971, p. 2) coins the term explicitly in reference to the term 'exobiology' used in Soviet research at the time.[2] The latter poses the question of the emergence and spread of life outside the earth, exosociology, on the other hand, asks accordingly about the emergence and spread of *intelligent* life—and about the possibility or impossibility[3] of communicating with extraterrestrial intelligences.

Apparently unaware of this Soviet debate (at least Kaplan's volume is not mentioned once), the concept of exo-sociology was reintroduced into scholarly discussion more than ten years later: In November 1983, a small English-language journal (*Free Inquiry in Creative Sociology*) published the essay "Towards an Exo-Sociology. Construction of the Alien" by Jan H. Mejer, a sociologist then teaching in Hawaii. The aim of the author, about whom little is known today,[4] was to establish a new *subfield of sociology* that would deal primarily with the question of how *strangeness* was and is socially constructed—and what could be derived from this in the future for our understanding of extraterrestrial civilizations. Mejer assigned a central role to the study of cultural knowledge about human-alien contact, as found especially in science fiction. He also asked how human sociology could find out something about the constitution of extraterrestrial societies if these should actually be discovered one day. From such questions Mejer expected impulses for a renewal of social and cultural scientific thinking even *before* the corresponding discovery.

Mejer's idea of a new sociological hyphen discipline never took hold in this form, and the program he formulated has not been realized—until today. The

---

[2] In the English-speaking world, however, the term 'astrobiology' has become established (until the 1990s, the term 'bioastronomy' was sometimes used).

[3] In contrast to SETI research of the Western type, the Soviet debate, not least under the influence of linguists such as B. V. Sukhotin, is characterized by a rather critical assessment of the possibilities of understanding (we will return to this in Chap. 4).

[4] When his essay on exosociology appeared, Jan Mejer was teaching at the *University of Hawai'i* (Hilo). A few years later, he went to Whitman College in Washington State, where he died in the late 1990s. At Whitman College, Mejer, whose sociological dissertation was on the "Theory of Disruptive Crisis," taught primarily in the field of social science environmental studies—there is still a "Jan Mejer Award for Best Essay in Environmental Studies" in his honor there. (We thank Professor Alton Okinaka of the University of Hawai'i and Professor Keith Farrington of Withman College for their insightful information on the work of Jan Mejer; personal correspondence with M. Schetsche).

author had posed questions that, at the time, were considered by most social researchers to be as scientifically meaningless as they were intellectually unproductive. And his basic idea was—this is probably the most important reason for the failure of this initiative—highly undesirable socio-politically at the beginning of the 1980s (more on this in a moment). Thus, over time, Mejer's theses, questions, and ideas were largely forgotten.

One of Mejer's few successors in terms of sociological thinking about space can today be regarded as the US sociologist Jim Pass with his "Astrosociology Research Institute"[5]—a project, however, that addresses much more earthly questions than those formulated by Mejer at the time. Jim Pass has obviously learned from the failure of his idealistic predecessor: at the centre of his work are processes of the social organisation of space research, namely manned space flight. Questions about the opportunities and risks of humanity's contact with extraterrestrial civilizations and the answers to them (whether literary or scientific), on the other hand, remain marginal topics.

The *actual task of an exosociology*,[6] as it was understood by Jan Mejer, but also by the Soviet researchers around Kaplan, namely the search for and the exploration of extraterrestrial civilizations, seems, on the other hand, to have faded to a small footnote in the history of sociology, a footnote that hardly seems worth mentioning in view of all the terrestrial problems we are confronted with today. Nevertheless, we have made it our goal to snatch this idea of an exosociology from oblivion. For we are firmly convinced that there are a number of good reasons—for sociology, for science, but also for society as a whole—not to take a hasty departure from those questions that Kaplan and Mejer each raised in their own time. Possibly the authors failed not because their ideas were nonsensical, but simply because they had formulated them at the wrong time— one could also say far too early. We will explain this only briefly using the example of what we call the *Hawaiian initiative*.

The period in which Mejer's programmatic essay appeared was not only the heyday of the so-called 'new social movements', which in all Western societies very vociferously promoted hitherto hidden issues onto the socio-political agenda (such as ecological dangers or the structural violence prevailing in many states), it was at the same time the end of the phase of a competitive but primarily peaceful exploration of space. In March 1983, half a year before Mejer's article

---

[5] See http://www.astrosociology.org/aboutARI.html.

[6] Under the heading of 'cosmos sociology', a quite similar programme is negotiated in the SF novel *The Dark Forest* by the bestselling Chinese author Cixin Liu (2018; Chinese original 2008). There, only the results of this discipline ultimately save humanity from malicious alien invaders—something that is probably only conceivable in science fiction.

appeared, the then US President Ronald Reagan had announced his vision of a sustained militarization of space, which was later to go down in history as the "Star Wars" program. The idea of space exploration guided by scientific interests—including possible *peaceful* contact with extraterrestrial civilizations—had been countered, at least in the US, by military strategic thinking.[7] In the early 1980s, civilian space exploration was to be subordinated to the primacy of space weapons research. Today we know that Reagan's plans were doomed to failure for a variety of reasons (not least technological). Nevertheless, the destructive positioning of the US administration poisoned the entire space research community for many years. In the social sciences (which at that time still tended to see themselves as 'critical of society'), this only had to reinforce the impression that the study of 'extraterrestrials' was morally acceptable only as part of a *cultural-critical* programme[8] to deconstruct violence-based power politics and to expose the machinations of the 'military-industrial complex'.

While large parts of the cultural and social sciences (as must be admitted here: often for good reasons) have held on to their fundamental condemnation of any not a priori ideology-critically oriented preoccupation with space and space research until today, the basic positions, empirical findings and also the visions of natural scientific space research have been able to develop further in many respects. In particular, in the last decade, the question of extraterrestrial civilizations, which moved Mejer, has also emancipated itself from fictional-literary thinking and has become a topic of scientific debate. After various discoveries in astro-science over the last three decades (for example, concerning the frequency of extrasolar planets, the widespread distribution of building blocks of life in space, or the extreme resilience of terrestrial organisms), the idea of life beyond Earth is now part of mainstream thinking in (natural) science. In the USA, *astrobiology* has even established itself as a new branch of research, which not only attracts the attention of the public and private and governmental sponsors with numerous innovative programmes and projects, but also exerts an increasing attraction on neighbouring disciplines (such as molecular biology or geology).

We will discuss this context in more detail in Chap. 3. Here it must suffice to point out that in the natural sciences today it is assumed that in the vastness of the universe there are not only in all probability numerous inhabited planets,

---

[7] It should not be forgotten, however, that space exploration, including manned space exploration, has always had a military and power-political dimension from its beginnings—but this is not the subject here (Welck 1986 and Schetsche 2005).

[8] We cannot systematically pursue this here, and therefore refer only by way of example to the volume by Günter Anders *Der Blick vom Mond* (1970), which gave a prominent voice to the cultural-scientific critique of space travel.

but that one could soon find evidence for the existence of those "extraterrestrial intelligent societies" of which Mejer spoke … if one only searched for them with the right methods. According to this assumption, under the keyword SETI (= Search for Extraterrestrial Intelligence) more and more new research programs have been established, which are committed to this goal: By receiving clearly artificial electromagnetic signals from the far reaches of the universe, they want to prove that there are other intelligent (and technically advanced) civilizations besides mankind (Engelbrecht 2008; Zaun 2010; Shuch 2011).

The actors of the various SETI campaigns share a body of basic scientific beliefs, which we have elsewhere (Schetsche and Engelbrecht 2008) called the "cosmic contact paradigm". It is based on a set of basic assumptions, building on each other, that (should) justify why the search for extraterrestrial intelligences makes scientific sense. Even if these assumptions are without exception epistemologically precarious (because they are all empirically based on the only known case of the development of life so far, namely that of the Earth itself), they meet with much approval today, not only among astro-scientists, but also in the scientific community in general. At least they are regarded with a certain goodwill—even if the optimism of the active SETI researchers regarding a quick success of the search is not always shared.

1.2 Tasks and Aims of Exosociology

The social framework conditions for a study of extraterrestrials, which is open in both senses of the word, are more favourable today than they have been for a long time, both in terms of the history of science and in cultural terms. And if one asks quite specifically for scientific reasons for a 're-animation' of the programme of exosociology, at least four arguments could be put forward:

Firstly: there are good astrophysical and astrobiological reasons for the assumption that humanity will sooner or later come into contact with extraterrestrial civilizations. And in all likelihood these civilizations will be at least technically far superior to our human ones (Shostak 1998, p. 121). The question of what such contact would mean for our terrestrial societies has long been simply ignored by SETI researchers themselves, as well as in cultural and social science. If we grant the SETI contact paradigm even a certain initial plausibility, *now* would be the time—in view of the increasing number of technically elaborate SETI projects and their active offshoots[9]—to seriously consider the possible

---

[9] Under the acronym METI (= Messaging to Extra-Terrestrial Intelligence) attempts have been made for years to draw the attention of extraterrestrials to Earth or the existence of

consequences of human-alien contact and to develop, within the framework of social science prognostics, appropriate scenarios for a 'case of the cases' that *could* become reality in the coming decades.

Since dealing with events or developments that are imaginable but not yet concretely foreseeable has not exactly been one of sociology's preferred fields of work in recent decades (cf. the discussions in the volume by Hitzler and Pfadenhauer 2005), however, additional arguments are needed to justify why it makes sense to put exosociological questions back on the scientific agenda today in particular.

Secondly: hardly any other topic points us more strongly to the complex set of conditions between reality-based fictional thinking in the present than the question of the relationship between humans and extraterrestrials. Since the Renaissance.

Fictional and speculative thinking about human-alien encounters provided countless scientifically usable ideas and thought experiments [...] and made this topic conceivable and explorable in the scientific context (for example in the form of SETI research) in the first place. On the other hand, scientific space research provided a variety of ideas, scenarios and background information for fictional formats: The observation of solar and extrasolar planets, theories on the origin and spread of life, and so on. (Schetsche & Engelbrecht 2008, p. 267).[10]

The themes of exosociology, newly brought into the scientific arena, thus represent an ideal (intellectual) experimental space for investigating the interactions between reality-based and artistic-literary thinking of our epoch (Grazier and Cass 2015). Not only do various processes of the 'social construction of reality' (Berger and Luckmann 1966) become particularly apparent here, but we are also confronted with the manifold variants of documents (and forms of thought) with *hybrid reality status that* increasingly dominate mass and network media today. A preoccupation with the 'human-alien problem' is at the same time always also a preoccupation with the question of what is considered reality in our society and what is not. (This is certainly most obvious in the so-called UFO question, which is anything but trivial scientifically—we will go into this in Chap. 11).

Thirdly: The theoretical preoccupation with the extraterrestrial as a prime example of a 'maximum stranger' (see Schetsche 2004; Schetsche et al. 2009) can prove to be a key issue for research into foreignness and xenophobia—which

---

mankind by means of their own radio messages—we will go into more detail about these experiments, which in our opinion are highly problematic, in Chap. 6.

[10] German quotes were translated into English for this book.

is urgently needed in socio-political terms. Here, a whole series of scientifically and socio-ethically significant questions arise: How is communication able to transcend linguistic and cultural boundaries? How can we think ourselves into a counterpart with a different world view or different modes of perceiving the world? What misunderstandings are to be expected when communicating with strangers? How can these be avoided if necessary? And, of course, even more fundamentally: How alien may a being be, so that we consider it to have equal rights—and grant it, for example, the legal status of a person? It is precisely the last question, which is ethically extremely serious, that necessarily provokes social, moral and scientific discussions about the determination of boundaries—such as those of human rights. From what point in time do they apply to the foetus? Until when for the dying? Under what conditions for apes? And when can an AI-controlled robot claim them? The alien example, which is still fictional today, helps us here to think fundamentally about the definition of intelligent life (which is therefore particularly worthy of protection according to the ethical standards that have dominated up to now): What constitutes an intelligent living being at all? Planned and intentional action? (What does it mean here if a living being is incapable of action?) Or the existence of a self-reflective consciousness? (What does this mean for coma patients?) Perhaps also the existence of a personality common in humans or a certain intelligence? Who determines this and by what means? These are all questions for which the exosociological thought experiments help us, because they enable us to search for answers to these and similar questions beyond the horizon of supposed everyday certainties. The theoretical study of extraterrestrials can thus also become a very practical touchstone for earthly moral judgments.

Fourthly: a basic problem with which every comparative cultural and social research is repeatedly confronted is the question of the significance of anthropological constants for the formation and self-structuring of societies. This corresponds directly with the (also socio-politically) highly controversial question of the interplay of biological predispositions and socializing influences in the development of individuals and entire cultures (Neyer and Spinath 2008). This question has not yet been answered satisfactorily because all experiments, studies and theoretical designs—if we disregard some borderline forms in other primates—are based on the investigation of a single culture-forming species, namely our own species. On the narrow basis of $N = 1$ it is simply not possible to decide between some competing hypotheses. This applies not only to the question of the relationship between biological and sociological factors in the development of culture, but also to a whole series of other problems in comparative cultural research (such as the general role of specific environmental factors, the significance of sensory perception for the formation of intelligence, or the necessity of an innate categorical or at least precategorical apparatus—the latter has been the subject of enduring philosophical and epistemological debate since Kant). All these complexes of questions are already today culturally emphatically thematized on the basis of fictional-hypothetical cases (as science fiction delivers them)—a real contact with an extraterrestrial culture would let completely new answer possibilities appear on the horizon—and this event would also generate numerous new questions. Not only for an exosociology that would then, at the latest, be forced to establish itself.

1.3   The Two Pillars of Exosociology

If one attempted to establish such a new hyphenated sociology, it would probably begin its journey into the scientific world with two legs and a cane. The cane would be the equally neighbouring and competing sociology of space (or 'astrosociology' in Jim Pass's sense)—we will not pursue that here.[11] The two legs, however, that our new discipline desperately needs to get going are, on the one hand, the sociology *of strangeness* and, on the other, *scientific futurology*. Let's start with the latter.

1.3.1   Exosociology as a Temporary Subfield of Futurology

Exosociology is not per se, but nevertheless today and for the foreseeable *future*, a primarily *futurological discipline*. This is due to the fact that we have not yet discovered an extraterrestrial civilization,[12] where we can ask: How is this civilization organized, how can we enter into 'conversation' with it (whatever that means here), and what impact would this discovery have on our human civilization? *Prior to* actual first contact with extraterrestrials, this latter point is undoubtedly at the center of exosociological interest, certainly does not constitute their only question, but probably their most important one. We look at how this question can be answered futurologically in detail in Chaps. 7 (preliminary

---

[11] For an introduction to the sociology of space, see Marsiske (2005) and the recent volume by Spreen and Fischer (2014).

[12] We discuss contrary assumptions in lay research, namely in the context of the so-called paleo-SETI tradition, in Chap. 11.

methodological considerations) and 8 (a multiple scenario analysis of first contact) of this book. At this point, a brief sketch of the futurological context in which such a question is embedded will suffice. To this end, let us begin with some general notes on the history of futurology as a social science discipline.

One of the abilities of humans that distinguishes them from most animals (such as rabbits, grasshoppers and lobsters—in the case of apes, dolphins and crows, on the other hand, we are uncertain due to the state of research[13]) is the ability to form a picture of the possible future and to develop corresponding plans of action. Corresponding to this is the fervent desire to gain certain knowledge about the future. This desire is revealed by the divination methods developed and practiced for thousands of years in almost all cultures: Divination, prophecy, future vision—be it with the help of the entrails of goats, by means of psychoactive substances or with rune stones (for examples see Hogrebe 2005; von Stuckrad 2007; Maul 2013).

Such procedures and experiences of the pre-modern era are followed in the modern era by attempts of philosophy and science to predict the future more or less reliably with their methods. This begins with the utopias of the Renaissance, which, however, are usually still a mixture of prediction and wishful thinking. In many cases, the argument remains unclear: *Will the* future look as described—or *should* it become so? Such a link can also be found, albeit with a much stronger guiding effect, in Marxist theory and utopia since the middle of the nineteenth century: socialism is coming—so the certain prognosis, but it "does not come like the dawn after a good night's sleep" (as Bertolt Brecht remarked). In other words, it is a desirable possibility, but whether or not it becomes a reality depends on the actions of particular actors. This variant of thinking about the future goes more clearly in the direction of scientific futurology as we know it today.

Futurology, which is scientific in the narrower sense because it is based on empirical data from the past, theories on the development of cultures and scientific methods of extrapolating time series, etc., emerged in the middle of the twentieth century. Especially in the USA, methods such as game theory, mathematical modelling, cybernetics, simulation techniques, the Delphi method, scenario techniques and many more were developed. So-called 'think tanks' in the environment of the military-industrial-political complex played and still play a central role in the USA. They were and are concerned with forecasting the development of their own and other societies against the background of power-strategic, often directly military-political considerations (Graf 2003; Kreibich 2006).

---

[13] See, for example, the contributions in Wirth et al. (2016).

But there is also a completely different tradition of futurology: as early as the 1940s, the German political scientist Ossip K. Flechtheim, a convinced pacifist who had been driven out of Germany by the Nazis, asked about the possibility of foreseeing social, political and military conflicts by systematically forecasting possible future developments—and of averting the realisation of such a reality through timely political action. In other words, futurology to avoid wars, not to make them waged and winnable by means of 'war games'. Flechtheim also coined the term futurology as the study of *possible futures*. This tradition of futurology experienced its heyday in the 1970s. During this period, three volumes were published within a few years that must still be regarded as tradition-building classics of this non-militaristic strand of futurology. It began in 1970 with the first edition of Ossip K. Flechtheim's *Futurologie* (Futurology). The subtitle *Der Kampf um die Zukunft* (The Struggle for the Future) already makes clear the essential thrust of the volume: it is not only about forecasting what the future might bring, but also about deciding on one of the possible futures—and then fighting politically for its realization. Three years later (1973), Robert Jungk published *Der Jahrtausendmensch* (The millennium man). Like Flechtheim, the author, a trained philosopher, had to flee Germany in the mid-1930s. He worked as a journalist in Switzerland throughout the war years, then worked as a foreign correspondent in many places around the world, and finally moved to Austria. There, in the late 1950s, he began to concern himself with the question of forecasting and planning for the future, and in particular with the problem of 'achieving' a *humane future for* all.[14] Almost exactly between these two volumes, the book was published in early 1972 from which, in the end, probably the strongest impulses for today's futurology, but also for the entire new, namely ecological, thinking in politics and society, have emanated: *The Limits to Growth* (Meadows 1972). The authors acted merely as representatives of the *Club of Rome,* founded in 1968 by intellectuals, scientists and humanists, which set itself the goal of fathoming the ecological problems facing humanity in the broadest sense and submitting proposals for their solution. At first glance, the special feature of this volume is that mathematical models for extrapolating time series are here applied excessively to *global* developments for the first time: World population, energy sources and raw materials, food, pollution. Only at a second glance, however, does it become apparent what constituted the real explosive power of this volume and made it one of the triggers of the so-called ecological revolution: the critique of the economic logic of growth and the introduction of an alternative idea of circulation

---

[14] Jungk was one of the masterminds of the peace and ecology movement. His book *Der Atomstaat* (1977) shaped the struggle against the use of nuclear energy for years.

and equilibrium. We have reported all this in comparative detail here because this is precisely the tradition[15] which exosociology, as a sub-field of futurology, explicitly claims for itself: to avert damage through its prognoses not primarily from individual nation states, but from humanity as such.

For futurology, the work of the Club of Rome initially brought a surge of new, computer-based procedures—and thus a significant methodological advance in *quantitative* forecasting. Virtually all computer-based 'world models' implemented later can be traced back to the methods developed here. Moreover, and this is particularly important for our context, the book highlights the importance of the analysis of *alternative scenarios*, an idea that has decisively shaped both quantitative and qualitative futures research to this day. Since the 1970s, futurology has regularly forecast not just one but *several possible futures*—and has attempted to indicate under which conditions one possibility and under which conditions the other could become reality.

We follow up this last point in Chaps. 8 and 9 of this book. We are convinced that the most important *current guiding question in exosociology* is: What would change for terrestrial societies if we gained the certain knowledge that humanity, as an intelligent species, is not alone in the universe? This is, since the event of the so-called 'first contact' has not yet taken place, for the time being still a *hypothetical* question—and thus one that is repeatedly posed in futurology. Within the framework of a futurological program, this first contact represents a so-called *wild card event*. Such events are characterized by the fact that, although the probability of their occurrence is low, if they do occur they are likely to lead to significant consequences that massively affect individual or a large number of subsystems of society (see Steinmüller and Steinmüller 2004). What the author pair Steinmüller has written about the consequences of many global[16] wild cards would certainly apply to first contact: Their effects "sometimes take on an almost fatal dimension" (p. 38), they have "the power to trigger shock waves of change" (p. 13).

---

[15] In this context it also belongs that the practical focus of Jan Mejer's work at the University of Hawaii was disaster research (personal communication from his former colleague, Professor Alton M. Okinaka—email to M. Schetsche dated 21.11.2017).

[16] First contact with an extraterrestrial civilization undoubtedly represents a global event with corresponding effects in the information-networked world. Such a contact in historical times, namely in human prehistory or early history, on the other hand, could well be conceived as a local event, affecting only the population of a narrowly defined region (this is the subject of paleo-SETI research, which we deal with in Chap. 11).

More on all this in Chap. 8 of this book. What is crucial at this point is that exosociology, with its question of 'what if', not only stands paradigmatically in the tradition of scientific futurology, but also moves methodologically and methodologically entirely within its framework—specifically in the form of *scenario analysis* (sometimes also called scenario technique), which is one of the most important methods of contemporary futurology (so Kosow and Gaßner 2008). Chapter 8 of this book reports in detail on the results of such a scenario analysis we conducted concerning the possible consequences of humanity's confrontation with an extraterrestrial civilization.

1.3.2 Exosociology as Part of a Sociology of Strangeness

The second stand and, more importantly, running leg of the exosociology we project is the *sociology of strangeness*, not only but especially in the form of the theoretical concept of the *maximal stranger* formulated years ago by one of us (Schetsche 2004; Schetsche et al. 2009) We cannot present the concept in detail here, but only sketch it out very roughly.

The stranger (in the sense of: foreign person) is regularly defined twice in social and cultural studies to this day: The stranger in the social (or everyday) sense is the one who is not personally known to me or does not belong to my social group. The stranger in the cultural (or structural) sense, on the other hand, is the one with whom I do *not* share the certainties that determine my world view. In his book *Topographie des Fremden* (topography of the foreign), Bernhard Waldenfels seems to introduce an additional, more borderline level: "radical strangeness. "However, a closer look reveals that this category is not directed towards a personal counterpart, but towards borderline phenomena of human existence, such as "eros, intoxication, sleep or death" (Waldenfels 1997, p. 37). Here the author addresses experiences which, although they transcend the everyday horizon of human meaning, nevertheless belong to the realm of human experience. If, on the other hand, the realm of the human lifeworld is left behind, then, according to Waldenfels, we are dealing with the "utterly alien", which simply cannot be investigated by the cultural and social sciences.

Georg Simmel could serve as a historical guarantor of such a scientific *exclusion*. In his classic excursus on the stranger from 1908, he stated quite apodictically: "The inhabitants of Sirius are not actually strangers to us—at least not in the sociologically relevant sense of the word—but they do *not exist for us at all*, they stand beyond far and near" (Simmel 1958, p. 509; emphasis by the authors). Such a restriction of reflection to the human being, which means—also quite practically—the exclusion of all types of 'non-humans' from scientific analysis, can be considered paradigmatic for the formation of theory in the cultural and social sciences until today.[17] The borderline of this conventional sociological concept of foreignness is traced by Vossenkuhl (1990, p. 109):

> If I recognize the foreign in myself, I will probably be more open to the foreignness of the other. But this openness and respect for the foreign are not sufficient for understanding the foreign. If I have understood my strangeness, I need not yet have understood that of the other or of another culture. On the contrary, I can deceive myself and uncritically project the foreign experience of myself onto the Other and his foreignness. The complement thesis suggests such a projection, or at least does not exclude it. [...] We encounter here a variant of the problem of the foreign-psychic: what entitles us to assume that the other feels and thinks about any things and events as we do; and what entitles us to assume that our understanding of the foreign other is identical with his understanding of himself?

And this, it should be added, applies all the more when it is a matter of a *non-human counterpart*. The alterity hypothesis constitutive of sociological action theory (see Knoblauch and Schnettler 2004) reaches its limit here, indeed it must fail from the outset (Schetsche et al. 2009, pp. 475–478). In this sense, the category of the maximally foreign approaches asymptotically the "borderline case of definite incomprehensibility" (Münkler and Ladwig 1997) standing at the very end of the continuum of familiarity, but without reaching it. Only beyond this borderline does *the* stranger transform into the foreign (Stagl 1997, p. 86), to which no status of an actor can be attributed. Between the outermost Limes of cultural strangeness and this border point, however, we are dealing, according to our initial thesis at the time, with the *maximally foreign*.

Speaking quite abstractly: The 'maximum stranger' categorically designates a counterpart who, according to the situational definition of the human actors involved,[18] is non-human, but nevertheless accepted in his subject status and addressed as at least potentially an equal partner in interaction. In this context,

---

[17] There are, however, exceptions. Stagl (1981, p. 279), for example, unfolds his argument explicitly in demarcation from Simmel's programmatic reference to the "inhabitants of Sirius". For him, science's mandate to investigate the possibilities and limits of interaction or communication explicitly applies to "all human beings or all rational beings" (p. 279)—even if the text does not address what kind of "rational beings" beyond human beings might be involved (cf. on this also Stichweh 1997, p. 165).

[18] Here it is assumed that the ontological as well as the communicative status of the respective actors is determined situationally by the participants and only then reproduced in scientific observation. If the scientific observer is himself a human being, the lifeworld reference of his reconstructions will be the situational definition of the human actors involved—not least

the maximum stranger appears to be a delimitable and at least potentially identifiable entity that—from a human point of view—has (a) a partial compatibility
of sensory and communication channels, (b) an at least rudimentary instance of
thinking and decision-making, (c) some form of self-awareness, (d) intentional
possibilities of action, and (e) a willingness to communicate in principle. At
the same time, the extent of the person's actual interactions with the maximal
stranger may vary considerably; moreover, the many qualities of the counterpart
may initially or even permanently remain uncertain (Stagl 1981, p. 279; Stichweh 1997, p. 165). Empirically turned, those maximal strangers include all those
non-human beings with whom interactions can take place, or at least seem (if
perhaps only hypothetically) possible: Domestic, farm and wild animals, gods,
angels and demons, artificial intelligences of all kinds, whether incorporated or
not—and precisely those extraterrestrial intelligences we are concerned with in
this volume.

It should be noted here that the category of the maximum stranger marks
the outermost area of what is conceivable and realizable in practice as a communicative counterpart and interaction partner in situations defined as social. A
meeting involving one or more such maximum strangers is constituted as a communicative borderline situation, in which a large part of those certainties which
we (can) unquestioningly take as a basis in all interactions between people are
dropped. The following list (we take it slightly modified from Schetsche et al.
2009) of such basic conditions of human ability to interact, which can be assumed
anthropologically with full justification, marks the problem lines of an interaction
with the maximum stranger. These basic assumptions must be critically *questioned in* advance in each concrete individual case—such as the encounter with
an extraterrestrial intelligence:

1. the biological origin and corporeality with a correspondingly environmentally
   adapted perceptual apparatus;[19]
2. the existence of vital physical needs (eating, drinking, sleeping, etc.);
3. the compatibility of sensory and communication channels as well as specific
   environmental representation and modes of world perception;
4. a commensurable ability to perceive shapes in spatial and spatio-temporal
   terms, as well as a similar temporal resolution of the perceptual apparatus;

---

because the corresponding definitions of the non-human participants will not be accessible
to him in advance.

[19] We will ask in Chap. 10 whether it is not much more likely that we will one day be
confronted with a post-biological 'machine civilization'.

5. the fact of natality and mortality and the knowledge of it;
6. a 'social nature', i.e. the dependence on other beings of the same species as well as the basic ability and willingness to communicate that necessarily goes along with it;
7. an openness to the world due to the detachment from biologically predetermined 'programs of thought and action';
8. concrete as well as abstract knowledge about the boundary between the self and the other, including the existence of a corresponding sense of self;
9. the capacity for self-interpretation or self-reflexivity.

From an exosociological point of view, this enumeration makes clear the central presuppositions of any engagement with extraterrestrial intelligences: *extraterrestrials are at most strangers*, with whom a large part of those basic certainties are omitted that we may unquestionably assume in interaction with human beings. In contact with an alien intelligence, they represent *inadmissible and avoidable* anthropocentric presuppositions that, at best, lead us to fail to understand our counterpart, and, at worst, generate misunderstandings that can prove highly disastrous for all involved. (We explore how such presuppositions enter traditional SETI research and complicate its work in Chap. 4).

1.4   Summary: Aims of Exosociology

At the end of this introduction we do not want to conceal this: It is all too legitimate to ask whether, in view of the numerous terrestrial problems presently demanding sociological attention, mental or even material resources should be devoted to exosociological considerations, studies, and experiments even before actual contact with an extraterrestrial civilization. Mejer, in his day, might have been quite unequivocal in affirming this—not least by pointing to the thought-expanding power of thinking about what is possible in principle. In our view, however, it is in particular the knowledge gained in the last twenty years about the nature of the universe in which we live that argues for such a professional experiment at precisely the present time: we now know that our cosmic environment is teeming with places potentially conducive to life—and that, on Earth at least, the first life arose almost as soon as external (physical and chemical) conditions permitted it. (We address these issues in detail in Chap. 3.) No one today knows with certainty whether life arose outside the Earth as well—much less whether other intelligent beings evolved in the vastness of space. But there is absolutely no reason to rule out this possibility. And given the sheer inconceivable size of the universe, it seems very likely (not only to us) that a multitude of highly alien civilizations exist alongside the Earthly civilization. And the more we know about the universe and the further we penetrate into the cosmos through our own research activities, the more likely it becomes that we will be confronted with those civilizations, their signals or legacies.

For these reasons, it seems sensible to us to present today, almost 35 years after Jan Mejer's forward-looking sketch, a renewed programmatic for exosociology. This sub-discipline at the crossroads of futurology and strangeness research should, against the background of current developments and expected future trends in the study of space, focus on five central thematic fields or guiding questions:

1. *The critical accompaniment of the scientific search for extraterrestrials*: How should the presuppositions of today's SETI, SETA and METI projects be assessed from a sociological perspective, and what contribution can the social sciences generally make to the further development of such research programmes?

2. *Interspecies futurology*: What would be the predictable consequences (the cultural and religious, the political and economic) of a human encounter with an extraterrestrial civilization? The answer to this question goes hand in hand with the further development of futurological methods, especially as far as the analysis of wild card events is concerned.

3. *Competing levels of reality*: What is the connection between scientific and fictional thinking about man's place in the cosmos and, in particular, the relationship between terrestrial and extraterrestrial civilizations?

4. *Research on foreignness and xenophobia*: How is foreignness culturally constructed today and what are the social and ethical consequences of the respective constructions—on earth and, in this case, also in encounters between the stars?

5. *Extra-human ethics*: What characteristics must a being possess, or how 'alien' may it be, so that we recognize in it an equal partner in interaction and grant it personal rights in principle? (Here, the points of connection to human-animal studies and robot ethics are immediately apparent).

It is clear to us that even a catalogue of questions expanded in this way will not change the fact that exosociology will remain a marginal field of cultural and social research for some time to come—at worst publicly ridiculed, at best scientifically tolerated. But this does not mean (as Jan Mejer already pointed out) that such a 'niche discipline' cannot provide impulses for the expansion or even renewal of scientific thought. And it certainly does not mean that its scientific marginal position must be permanent: at the latest with the proof of the existence of an intelligent species outside the earth, exosociology would move into the centre of professional *and* public interest. Ultimately, this might even make it a kind of leading discipline on call.

References

Anders, Günter. 1970. *Der Blick vom Mond. Reflexionen über Weltraumflüge.* München: C. H. Beck.

Berger, Peter L. & Thomas Luckmann. 1966. *The Social Construction of Reality.* New York: Doubleday.

Engelbrecht, Martin. 2008. SETI – Die wissenschaftliche Suche nach außerirdischer Intelligenz im Spannungsfeld divergierender Wirklichkeitskonzepte. In *Von Menschen und Außerirdischen. Transterrestrische Begegnungen im Spiegel der Kulturwissenschaften*, Hrsg. Michael Schetsche und Martin Engelbrecht, 205–226. Bielefeld: transcript.

Flechtheim, Ossip K. 1970. *Futurologie. Der Kampf um die Zukunft.* Köln: Verlag Wissenschaft und Politik.

Graf, Hans Georg. 2003. Was ist eigentlich Zukunftsforschung. *Sozialwissenschaft und Berufspraxis* 26 (4): 355-364.

Grazier, Kevin R. & Stephen Cass. 2015. *Hollyweird Science: From Quantum Quirks to the Multiverse.* New York: Springer.

Hitzler, Ronald & Michaela Pfadenhauer, Hrsg. 2005: *Gegenwärtige Zukünfte. Interpretative Beiträge zur sozialwissenschaftlichen Diagnose und Prognose.* Wiesbaden: VS Verlag.

Hogrebe, Wolfram, Hrsg. 2005. *Mantik. Profile prognostischen Wissens in Wissenschaft und Kultur.* Würzburg: Königshausen und Neumann.

Jungk, Robert. 1973. *Der Jahrtausendmensch. Bericht aus den Werkstätten der neuen Gesellschaft.* München: Bertelsmann.

Jungk, Robert. 1977. *Der Atom-Staat—Vom Fortschritt in die Unmenschlichkeit.* Stuttgart: Bücherbund.

Kaplan, S. A. 1971. Exosociology - the Search for Signals from Extraterrestrial Civilisations. In *Extraterrestrial Civilizations. Problems of Interstellar Communications*, ed. S. A. Kaplan, 1–12. Jerusalem: Israel Program for Scientific Translations (Russian 1969) .

Knoblauch, Hubert & Bernt Schnettler. 2004. „Postsozialität", Alterität und Alienität. In *Der maximal Fremde. Begegnungen mit dem Nichtmenschlichen und die Grenzen des Verstehens*, Hrsg. Michael Schetsche, 23–42. Würzburg: Ergon.

Kosow, Hannah & Robert Gaßner. 2008. Methoden der Zukunfts- und Szenarioanalyse – Überblick, Bewertung und Auswahlkriterien (IZT-Werkstattbericht 103). Berlin: Institut für Zukunftsstudien und Technologiebewertung. https://www.izt.de/fileadmin/publikationen/IZT_WB103.pdf.

Kreibich, Rolf. 2006. Zukunftsforschung (IZT-Arbeitsbericht 23/2006). http://www2.izt.de/pdfs/IZT_AB_23.pdf.

Marsiske, Hans-Arthur. 2005. *Heimat Weltall. Wohin soll die Raumfahrt führen?* Frankfurt am Main: Suhrkamp.

Maul, Stefan. 2013. *Die Wahrsagekunst im alten Orient.* München: Beck.

Meadows, Dennis et al. 1972. *The Limits to Growth. A Report for the Club of Rome's Project on the Predicament of Mankind.* New York: Universe Books

Mejer, Jan H. 1983. Towards an Exo-Sociology: Constructs of the Alien. *Free Inquiry in Creative Sociology* 11 (2): 171-174.

Münkler, Herfried & Bernd Ladwig. 1997. Dimensionen der Fremdheit. In *Furcht und Faszination. Facetten der Fremdheit*, Hrsg. Herfried Münkler, 11–43. Berlin: Akademie Verlag.

Neyer, Franz J. & Frank M. Spinath, Hrsg. 2008. *Anlage und Umwelt. Neue Perspektiven der Verhaltensgenetik und Evolutionspsychologie.* Stuttgart: Lucius & Lucius.

Schetsche, Michael, René Gründer, Gerhard Mayer & Ina Schmied-Knittel. 2009. Der maximal Fremde. Überlegungen zu einer transhumanen Handlungstheorie. *Berliner Journal für Soziologie* 19 (3): 469–491.

Schetsche, Michael. 2004. Der maximal Fremde - eine Hinführung. In *Der maximal Fremde. Begegnungen mit dem Nichtmenschlichen und die Grenzen des Verstehens*, Hrsg. Michael Schetsche, 13–21. Würzburg: Ergon.

Schetsche, Michael. 2005. Rücksturz zur Erde? Zur Legitimierung und Legitimität der bemannten Raumfahrt. In: *Rückkehr ins All* (Ausstellungskatalog, Kunsthalle Hamburg), 24–27. Ostfildern: Hatje Cantz.

Shostak, Seth. 1998. Sharing the Universe. Perspectives on Extraterrestrial Life. Berkeley: Berkeley Hills Books.

Shuch, H. Paul. 2011. *Searching for Extraterrestrial Intelligence - SETI Past, Present, and Future.* Berlin: Springer.

Simmel, Georg. (4. Aufl.) 1958. *Soziologie: Untersuchungen über die Formen der Vergesellschaftung.* Berlin: Duncker & Humblot.

Spreen, Dierk, und Joachim Fischer. 2014. *Soziologie der Weltraumfahrt.* Bielefeld: transcript.

Stagl, Justin. 1997. Grade der Fremdheit. In *Furcht und Faszination – Facetten der Fremdheit*, Hrsg. Herfried Münkler, 85-114. Berlin: Akademie Verlag.

Stagl, Justin. 1981. Die Beschreibung des Fremden in der Wissenschaft. In *Der Wissenschaftler und das Irrationale*, zweiter Band, Hrsg. Hans Peter Duerr, 273–295. Frankfurt am Main: Syndikat.

Steinmüller, Angela & Heinz Steinmüller. (2. Aufl.) 2004. *Wild Cards. Wenn das Unwahrscheinliche eintritt.* Hamburg: Murmann.

Stichweh, Rudolf. 1997. Ambivalenz, Indifferenz und die Soziologie des Fremden. In: *Ambivalenzen. Studien zum kulturtheoretischen und empirischen Gehalt einer Kategorie zur Erschließung des Unbestimmten.* Hrsg. Heinz Luthe & Rainer E. Wiedemann, 165-183. Opladen: Leske + Budrich.

Stuckrad, Kocku von. 2007. *Geschichte der Astrologie. Von den Anfängen bis zur Gegenwart.* München: Beck.

Vossenkuhl, Wilhelm. 1990. Jenseits des Vertrauten und Fremden. In *Einheit und Vielfalt.* XIV. Dt. Kongress für Philosophie Giessen, 21.-26. September 1987, Hrsg. Odo Marquard, 101–113. Hamburg: Meiner.

Waldenfels, Bernhard. 1997. *Topographie des Fremden. Studien zur Phänomenologie des Fremden*, Band 1. Frankfurt am Main: Suhrkamp.

Welck, Stephan Frhr. von 1986. Weltraum und Weltmacht. Überlegungen zu einer Kosmopolitik. *Europa-Archiv* 41 (1): 11–18.

Wirth, Sven et al. Hrsg. 2016. *Das Handeln der Tiere. Tierische Agency im Fokus der Human-Animal-Studies.* Bielefeld: transcript.

Zaun, Harald. 2010. *SETI – Die wissenschaftliche Suche nach außerirdischen Zivilisationen. Chancen, Perspektiven, Risiken.* Hannover: Heise.

Thinking About Aliens

2

Since time immemorial, gazing up at the starlit night sky has inspired mankind's imagination and thirst for knowledge. As early as the Stone Age, astronomical observations formed the basis for cultic worship of the stars, but also for knowledge about the seasons and the first calendar systems. The astronomical alignments of numerous prehistoric burial sites and places of worship—the megalithic stone circle structure at *Stonehenge* in southern England is certainly one of the most impressive structures of this kind—provide impressive evidence of the importance that early human cultures attached to the celestial phenomena they observed. The sun, moon, stars, and planets and the deities associated with them continued to form elemental components of ancient mythologies. Beyond mythical-religious ideas about the nature and workings of *supernatural* entities in the 'heavens', speculations about *extraterrestrial* life forms on the celestial bodies surrounding the earth also existed at least since antiquity. A fragment of the Pythagorean-Egyptian 'Orphic Songs', handed down by the Neoplatonist Proclus (412–485 A.D.), states that mountains, cities and proud buildings rise on the moon. The pre-Socratics Xenophanes of Colophon (b. c. 570 B.C.) and Philolaos of Kroton (b. c. 470 B.C.) taught that the moon was habitable. And Democritus (460–371 B.C.), the founder of the theory of the atom, also had astonishingly modern ideas about the origin, decay and habitability of alien worlds:

> Some are still growing, others are at the height of their bloom; others are on the wane [...]. In some there is neither sun nor moon, in some they are larger than in our world and in some there are more of them. [...] And there were some worlds in which there were no animals and plants and no moisture at all" – but some of these innumerable worlds "were not only similar to one another, but in every respect completely, indeed so completely, the same that there was no difference at all among them, and likewise it would be with the people there. (quoted from Oeser 2009, p. 16)

© The Author(s), under exclusive license to Springer Fachmedien Wiesbaden GmbH, part of Springer Nature 2023
A. Anton and M. Schetsche, *Meeting the Alien*,
https://doi.org/10.1007/978-3-658-41317-0_2

In his work *De facie in orbe lunae* (On the Moon's Face), the Greek philosopher Plutarch (b. ca. 45 A.D.) speculates on the Earth-like nature of the Moon and on possible inhabitants of the Earth's satellite. He assumes that the moon, just like the earth, has mountains, valleys, etc., which explain the impression of a 'face' on its surface—a thoroughly revolutionary view at the time, since the philosophical doctrine of the Aristotelian and Stoic schools held that the moon consists of condensed 'ether'. But since the surface structure of the moon is similar to that of the earth, Plutarch concludes, life should also be possible there: "One would have to believe that it was created without purpose and meaning if it did not produce fruit, provide a dwelling place for human beings, enable their birth and nourishment, things for the sake of which, according to our conviction, our earth was also created" (Plutarch 1968, p. 56).

2.1   Aliens in Modern Times

In the Middle Ages, which were shaped by the Aristotelian-Ptolemaic worldview and scholasticism, such ideas hardly played a role. It was not until the intellectual upheavals of the Renaissance and the accompanying overcoming of geocentrism that extraterrestrials once again became the subject of scientific-philosophical considerations: Theologians, philosophers and natural scientists such as Nicholas of Cusa, Giordano Bruno, Nicolaus Copernicus, Galileo Galilei and Johannes Kepler, but also the authors of the early utopian novels, for example Francis Godwin or John Wilkins, dealt with the question of the habitability of alien worlds (Heuser 2008, p. 55).

During the Renaissance, it gradually became accepted that the sun, and not the earth, was the centre of planetary motion. This was initially not primarily a result of astronomical observations, but of natural philosophical considerations. In his work *De docta ignorantia* from 1440—i.e. even before Copernicus—Nicholas of Cusa (1401–1464) argues that the Earth is not at the centre of the universe, that contrary to sensory perception it is not at rest but in motion, and that it is only one of countless worlds in an unlimited but not infinite universe. Von Kues was convinced that life does not exist only on Earth, but occurs in a wide variety of forms throughout the universe and may have reached higher levels of development elsewhere than Earth-bound organisms (Heuser 2008, pp. 59–62). He writes:

> Hence, since that entire region is unknown to us, those inhabitants remain altogether unknown. [...] We surmise that in the solar region there are inhabitants which

are more solar, brilliant, illustrious, and intellectual–being even more spiritlike than [those] on the moon, where [the inhabitants] are more moonlike, and than [those] on the earth, [where they are] more material and more solidified. [...] In like manner, we surmise that none of the other regions of the stars are empty of inhabitants – as if there were as many particular mondial parts of the one universe as there are stars, of which there is no number. (von Kues 1985, pp. 171–172)

In agreement with Nicholas of Cusa, Giordano Bruno (1548–1600) also assumed that there must be countless inhabited worlds in the universe. In his writing *De immenso*[1] (1591) Bruno develops comprehensive cosmological considerations, with which he was far ahead of his time. Rejecting the Aristotelian conception of a shell universe taught by medieval scholastics, Bruno describes an infinite universe with countless worlds as well as the universal nature of space and natural laws. Since identical basic conditions prevailed everywhere in the universe, it is a logical consequence for Bruno that other planets also bear life, which, depending on the concrete composition of the elementary basic substances, could assume the most varied forms. Some of Bruno's statements on life in space almost seem like postulates of modern astrobiology, for example when he writes: "Just as on the surface of the earth some living beings live in the sea, some in various other areas, so it must be imagined with every other star, where the same kinds of basic substances combine to form the composition of a heterogeneous whole" (quoted from Heuser 2008, p. 71). Elsewhere it is said that other worlds, "contain living beings and inhabitants as well as this world can, since they have neither lesser powers nor are of a different nature" (quoted from Akerma 2002, p. 56). It was precisely this assumption of a universe populated by countless extraterrestrial life forms that also fueled Bruno's conflict with the Church. From their point of view, the doctrine of the multiplicity of inhabited worlds fundamentally violated Christian beliefs:

[...] the theological centrality of man, which is related to sin, the incarnation of God in Jesus and redemption. In the case of the habitability of countless other worlds, the drama of creation and redemption no longer has only one axis, but – countless ones. The Christian dramaturgy is sensitively disturbed and overtaxed when the drama of creation and redemption is to be performed on multiple stages (planets). (Akerma 2002, p. 52)

---

[1] *De immenso* is a common abbreviation in literature and research for the full title of the writing: *De Innumerabilius, Immenso et Infigurabili, seu De Universo et Mundus libri octo* (Heuser 2008, p. 62).

After Bruno's arrest, the Church demanded a complete recantation of his teachings, but he persisted in his claim of countless inhabited worlds. On February 17, 1600, Giordano Bruno was executed at the stake. It can rightly be asked why a similar fate did not befall Nicholas of Cusa, who, as it were, anticipated Bruno's thoughts. An answer can be found in Guthke:

> If not Nicholas of Cusa, but Giordano Bruno, who was intellectually dependent on him, had to atone for this heresy at the stake, it was because by now Copernicus' *De revolutionibus orbium coelestium* (1543) had given the idea of the plurality of worlds an entirely different philosophical status – that of potential reality and secular truth. (Guthke 1983, p. 43; emphasis in original)

Galileo Galilei (1564–1642), who was a contemporary of Giordano Bruno and who, as is well known, also got into a conflict with the church (although less consequential for him), saw the ancient thesis of its similarity to the earth confirmed by his observations of the moon with telescopes. Galileo recognized rough structures, plains, depressions, elevations, etc., which reminded him of the surface of the earth, but emphasized that life on the moon, if present, must be of a completely different nature than life forms on earth:

> Imagine what the consequences would be if the hot zone were irradiated by the sun without interruption for half a month; it is understood that all trees, herbs, and animals would infallibly be destroyed. If, therefore, a production of life did take place on the moon, it could only be plants and animals of an entirely different nature. (quoted from Oeser 2009, p. 25)

Johannes Kepler (1571–1630) also took up the ancient thoughts on possible inhabitants of the moon, but went much further than Galileo. In his posthumously published narrative *Somnium, seu opus posthumum de astronomia lunari* (1634), Kepler mixed astronomical considerations with social utopian and magical narrative elements, yet the writing "can be considered the first scientific work on comparative celestial body science" (Heuser 2008, p. 73). Kepler saw in the structures on the lunar surface observable through the telescope constructions of immense scale, built by the lunar inhabitants to protect themselves against solar radiation and possibly also against enemies. From the size of these structures, Kepler deduced the size of the lunar inhabitants, which must exceed that of humans many times over (Oeser 2009, pp. 25–30). Kepler divided the lunar inhabitants, who were generally "serpentine", into two groups: He called the inhabitants of the Earth-facing side of the Moon *Subvolvanians*, and those of the

Earth-defacing side *Privolans*. Kepler fancifully described how the lunar inhabitants lay down in the sun during the day "for their pleasure", but only "very close to their burrows, so that they can retreat quickly and safely" (Günther 1898, p. 21)—similar to lizards and crocodiles. With the Privolans there are.

> No safe and fixed abode, the lunar creatures cross their whole world in droves during a single day, following the receding waters partly on foot, equipped with legs longer than our camels, partly with wings, partly by ship. [...] Most of them are divers, all of them are creatures that breathe very slowly by nature, so they can spend their lives deep at the bottom of the water, coming to nature's aid through art. (Günther 1898, p. 20).

The ideas and discoveries of Nicholas of Cusa, Nicolaus Copernicus, Giordano Bruno, Galileo Galilei and Johannes Kepler—even if they sometimes met with massive resistance at the time—permanently broadened the horizons of human thought and not only prepared the way for modern astronomy, but also inspired a new genre of narrative art: *science fiction*.[2] The first Anglo-Saxon science fiction narratives by Francis Godwin (1562–1633) and John Wilkins (1614–1672) were significantly influenced by the "transterrestrial philosophy" (Heuser 2008, p. 75) of Nicholas of Cusa and Giordano Bruno, as well as by Kepler's *Somnium narrative*. Subsequently, the new genre continued to draw on the thoughts of Renaissance scholars—all the way to Jules Verne's (1828–1905) famous 1870 *voyage around the moon, which* in turn influenced early space pioneers such as Hermann Oberth, Rudolf Nebel, and Wernher von Braun, who eventually technically realized the ancient human fantasy of a journey to the moon (Heuser 2008, pp. 76–77).

2.2 Locke and Kant on Extraterrestrials

The thinking about extraterrestrials that intensified during the Renaissance also continued in philosophy. For the Enlightenment philosopher John Locke (1632–1704), there was no doubt that extraterrestrial life existed—including forms superior to humans. In his 1690 work *An Essay Concerning Human Understanding,* Locke wrote, referring to our solar system, "What several sorts of vegetables,

---

[2] At that time, however, such stories were not yet called that. The term 'science fiction' was not coined until the middle of the nineteenth century. Nevertheless, the works of Godwin and Wilkins can be regarded as early science fiction narratives or as precursors of modern science fiction.

animals, and intellectual corporeal beings, infinitely different from those of our little spot of earth, may there probably be in the other planets [...] (Locke 1999, p. 546). This may at first seem surprising, since Locke was known to hold the principle that knowledge must be preceded by *sensory experience* ("Nihil est in intellectu quod non prius fuerit in sensibus"[3]), and there can admittedly be no question of sensory perception of extraterrestrials. Accordingly Locke adds:

> [...] to the knowledge of which, even of their outward figures and parts, we can no way attain whilst we are confined to this earth; there being no natural means, either by sensation or reflection, to convey their certain ideas into our minds? They are out of the reach of those inlets of all our knowledge: and what sorts of furniture and inhabitants those mansions contain in them we cannot so much as guess, much less have clear and distinct ideas of them. (Ibid.)

But what then leads Locke to the conviction that extraterrestrial life exists? For Locke, a proven means of gaining knowledge about areas that elude our sensory perception is the formation of analogies. Since the universe is gigantic, life on earth manifold, and the power and wisdom of God inexhaustible, it would be presumptuous to assume that the Earth is the only inhabited planet and humanity the most highly evolved form of life. Locke writes:

> He that will not set himself proudly at the top of all things, but will consider the immensity of this fabric, and the great variety that is to be found in this little and inconsiderable part of it which he has to do with, may be apt to think that, in other mansions of it, there may be other and different intelligent beings, of whose faculties he has as little knowledge or apprehension as a worm shut up in one drawer of a cabinet hath of the senses or understanding of a man; such variety and excellency being suitable to the wisdom and power of the Maker. (Locke 1999, p. 103)

The Dutch astronomer and mathematician Christiaan Huygens (1629–1695), who is regarded as the founder of the wave theory of light and constructed the first pendulum clocks, argued in a very similar way. In his treatise *Cosmotheoros,* published posthumously in 1698, he summarized the state of knowledge about the solar system at that time and also dealt in detail with the possible properties of the inhabitants of alien planets. To Huygens, it seemed nonsensical that God could have created the planets without life existing on them to marvel at the magnificence of God's creation. Huygens assumed that there must be innumerable different manifestations of extraterrestrial life and that these would depend on the particular conditions of their home planets. Thus animals on Jupiter or Saturn

---

[3] Translated, "Nothing is in the mind that was not first in the senses".

would be ten or fifteen times larger than elephants. Higher-evolved extraterrestrial beings, Huygens argues, would probably think about stars and planets just as humans do, and would develop astronomy, mathematics, art, and music (Moore 2014, p. 210). Like Locke, Huygens argues that there is no reason to assume that humanity is the most highly evolved life form. The aliens, he writes:

> [...] probably build themselves huts and houses, or dig out caves; and this is the more probable, because with us all animals except fishes build something of the kind for their abode. But why only huts and little houses: if we are not to believe that the inhabitants of the planet build large and magnificent houses, we must esteem our own things as much more beautiful and perfect. But why should it be better with us in particular? Perhaps it is because we live on this little sphere, which, compared with the spheres of Saturn and Jupiter, has not even the ten-thousandth part of the contents of the body. So also no reason can be given why on these planets the beauty of architecture and its symmetry should not be as well known as with us, nor why they do not build palaces, towers, or pyramids, which are perhaps here and there higher and grander than with us. (Huygens 1703, p. 63)

Immanuel Kant (1724–1804) also dealt with the subject of extraterrestrials several times in his work—and not just 'in passing' or as a footnote:

> Rather, Kant repeatedly comes to speak of aliens at quite central points of his thought, and this throughout his creative period, so that for this reason alone it would be wrong to dismiss Kant's exobiological interest as either a sin of youth or the confusion of age. (Wille 2005, p. 11)

Kant assumed that there were a multitude of inhabited planets and that the distance of the planets from their home star and their material composition were in proportion to the spiritual qualities of their inhabitants: A planet far from the sun must be composed of lighter matter than a planet near the sun, therefore its inhabitants must also be composed of more mobile, finer matter, and consequently possess a more active, more perfect mind than the inhabitants of planets near the sun. Specifically, the inhabitants of Mercury and Venus were to be regarded as spiritually inferior, but those of Jupiter and Saturn as spiritually superior (Akerma 2002, p. 167; Dick 1982, pp. 173–174). Interestingly, for Kant there was a fundamental categorical difference between belief in extraterrestrials and belief in spirit beings. Extraterrestrials, he argued, were in principle accessible to human senses, even though humanity would probably never get to see them because of the distances between the planets. In principle, however, according to Kant, the existence of extraterrestrials would be *empirically* provable (Hövelmann 2009, p. 178).[4]

The fact that some of the most important natural scientists and philosophers of the seventeenth and eighteenth centuries took up the question of the possibility of extraterrestrial intelligences can on the one hand be understood as a homage to their ancient predecessors and on the other hand is due to the strengthening of a science increasingly freed from dogmatic thinking by the intellectual upheavals of the Renaissance and the Enlightenment. According to Dick (1982, I), the question of extraterrestrial intelligences had thus made it into the orthodoxy of Western thought. One does not necessarily have to agree with this thesis, nevertheless it can be said that hypothetical, potentially rational extraterrestrials from now on at least (again) moved into the realm of the principally thinkable and became a (reasonably) accepted topic of discourse in Western societies.

2.3   Aliens in Modern Science Fiction

Back to science fiction: just as early science fiction, which was shaped by natural philosophical considerations about extraterrestrials, influenced the development of technology, from the end of the nineteenth century onwards technological development increasingly shaped thinking about extraterrestrials. The technological innovations of the time triggered optimism about progress, faith in science, and technologically inspired fantasies of redemption on the one hand, but also massive fears about the effects of technology on humans on the other. This ambivalent relationship to technology, which was believed to have the potential to realize utopian societies as well as the potential for the complete annihilation of mankind, formed (and still forms) a central motif of science fiction—and was also projected onto the fictional extraterrestrials. Henceforth, these aliens often possess highly advanced technology, which they direct against humanity with warlike intent in more than a few stories, such as the science fiction classic *War of the Worlds* by H. G. Wells from 1898, in which technologically advanced Martians attempt to conquer Earth with the aid of fearsome three-legged fighting machines in order to exploit its resources. The human military is hopelessly inferior to the aliens'

---

[4] It is different with ghosts: In his 1766 essay *Träume eines Geistersehers (Dreams of a Spirit-Seer)*, Kant argues that spiritual beings are not accessible to human experience and that their existence is therefore neither provable nor disprovable. He writes: "One can therefore assume the possibility of immaterial beings without fear of being refuted, though without hope of being able to prove this possibility by reason" (Kant 1954, p. 19).

superior technology. This motif of the (technological) superiority of the aliens is found right at the beginning of the novel, where it says:

> No one would have believed in the last years of the nineteenth century that this world was being watched keenly and closely by intelligences greater than man's and yet as mortal as his own; that as men busied themselves about their various concerns they were scrutinised and studied, perhaps almost as narrowly as a man with a microscope might scrutinise the transient creatures that swarm and multiply in a drop of water. With infinite complacency men went to and fro over this globe about their little affairs, serene in their assurance of their empire over matter. It is possible that the infusoria under the microscope do the same. No one gave a thought to the older worlds of space as sources of human danger, or thought of them only to dismiss the idea of life upon them as impossible or improbable. It is curious to recall some of the mental habits of those departed days. At most terrestrial men fancied there might be other men upon Mars, perhaps inferior to themselves and ready to welcome a missionary enterprise. Yet across the gulf of space, minds that are to our minds as ours are to those of the beasts that perish, intellects vast and cool and unsympathetic, regarded this earth with envious eyes, and slowly and surely drew their plans against us (Wells 1992).

After their arrival on Earth, the alien invaders mercilessly destroy major cities, communications and transportation networks. Only by luck can humanity escape their complete annihilation: Earthly bacteria infest and kill the alien invaders (Fig. 2.1).

The fact that Wells' extraterrestrial invaders are, of all things, Martians is no coincidence: around the turn of the century there was a veritable 'Mars boom' in science fiction literature, but also in scientific discourse, and the Western world was fascinated by the idea that Mars could harbour life—perhaps even intelligent life (Kaiser 2004). The Mars euphoria was triggered, among other things, by the 'Martian canals', geological structures on the surface of Mars first described in 1877 by the Italian astronomer Giovanni Schiaparelli (1835–1910), which were interpreted, in particular by the French astronomer Nicolas Camille Flammarion (1842–1925) in several highly circulated books, as gigantic constructions of a highly developed extraterrestrial civilization on Mars (Moore 2014, pp. 216–217). The 'canals' relatively quickly turned out to be natural structures, optical illusions, photo-technical effects and also deliberate deceptions. However, this did not dampen the general enthusiasm for the idea of life on Mars (Oeser 2009, pp. 165–168) (Fig. 2.2).[5]

---

[5] Something similar was repeated many decades later in the context of the discussions surrounding the so-called 'Martian face', a rock formation photographed in 1976 by the orbiter of the *Viking I* space probe, which in the first photograph resembles a human face and triggered speculation as to whether it might be the remains of an extraterrestrial high civilization on Mars (Hoagland 1987).

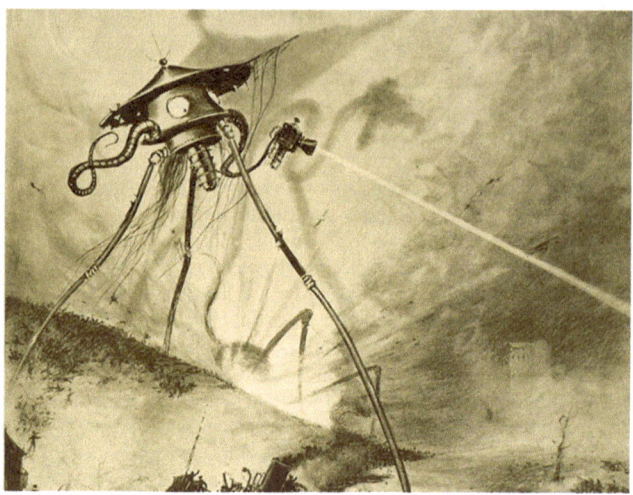

Fig. 2.1 Artist's rendering of the Martian invaders in *War of the Worlds* by Alvim Corréa from 1906. *Source* Public domain image[6]

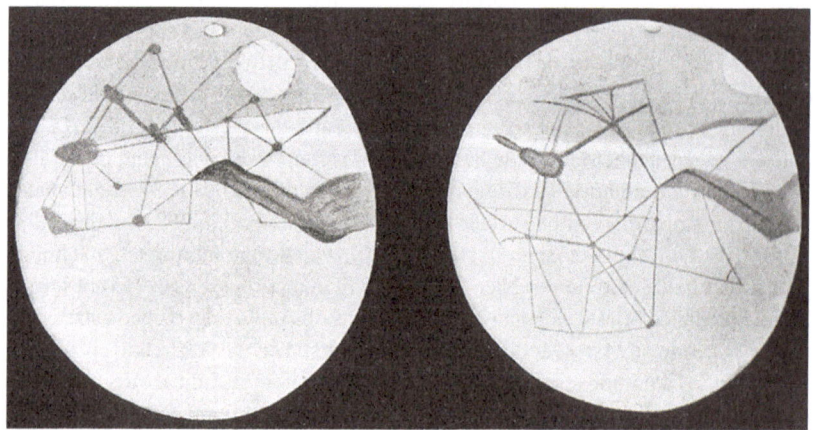

Fig. 2.2 Illustration of Percival Lowell's 'Martian Channels'. *Source* Public domain figure[7]

---

[6] Available online at: https://de.wikipedia.org/wiki/Der_Krieg_der_Welten#/media/File: War-of-the-worlds-tripod.jpg.

[7] Available online at: https://de.wikipedia.org/wiki/Der_Krieg_der_Welten#/media/File: War-of-the-worlds-tripod.jpg.

On October 30, 1938, the U.S. radio station CBS broadcast a radio play of *War of the Worlds,* done in the style of a live radio broadcast, in which a reporter reports the details of the alien invasion. Apparently, some listeners to the broadcast had missed the clues that this was a radio play, because they thought the alien invasion was real and panicked. That there had been a real mass panic in the course of the broadcast, as was reported in some newspapers the next day, turned out to be a strong exaggeration of the press afterwards, but nevertheless the broadcast had an enormous effect: Several police stations received so many calls from concerned citizens during and after the broadcast that at times the lines broke down, frightened people ran into the streets with wet clothes in front of their faces to protect themselves from the 'poison gas' of the Martians, and quite a few angry men inquired with the authorities where and how they could join the armed resistance against the alien invaders (Gerritzen 2016, pp. 13–20).

Both H. G. Wells' novel and the 1938 radio broadcast of *War of the Worlds* point to remarkable developments in the relationship between humans and aliens: The novel represents the tendency in science fiction literature to portray aliens increasingly as *acting agents who* communicate and interact with humans, guided by certain motivations and interests, and thus becoming more and more a projection screen for human experiences, longings, hopes, and fears. As a result, the fictional aliens became the 'vehicle' for narratives in which almost every conceivable collective human experience and emotion, the most diverse social designs, philosophical considerations, political currents, utopias and dystopias were negotiated. The described reactions to the radio play of *War of the Worlds,* in turn, mark a remarkable circumstance: apparently, the idea of the existence of extraterrestrial civilizations was widespread in the USA in 1938 and was taken so seriously that an attack by extraterrestrial invaders was considered a *real possibility* by at least some listeners to the CBS radio broadcast. Here, the interplay between fictional and reality-based thinking alluded to in the introduction is impressively revealed: the imagined aliens suddenly appeared as factuality, they penetrated the everyday reality of people as a real threat—even if only for a few hours.

Over the course of the twentieth century, aliens became an increasingly important (and sometimes highly profitable) element of science fiction narratives—today it's impossible to imagine the genre without them. In countless.

Books, films and television series of the science fiction genre feature encounters with extraterrestrial beings: What must remain speculation in reality becomes a meaningful experience in the imagination of creative writers and filmmakers and in the collective imagination of a worldwide audience. (Hurst 2008, p. 33)

The particular appeal of the extraterrestrial subject in the context of artistic-fictional narratives results from the fact that the figure of the extraterrestrial as the *maximum stranger* (Schetsche 2004; Schetsche et al. 2009) condenses all discursive fields of association, psychological patterns of perception and emotional modes of reaction in relation to an alien counterpart, which, combined with our factual ignorance about potential real extraterrestrials, open up an almost limitless space for creative creations. The ambivalence or contingency of encounters with strangers confronts us particularly effectively in our fantasies about extraterrestrial beings: "Strangers imply the absence of clarity, one cannot be sure what they will do, how they would react to one's actions; one cannot tell whether they are friends or enemies—and therefore one cannot help but regard them with suspicion" (Bauman 2000, p. 39). This uncertainty increases how much the stranger is a counterpart to us. Or, conversely, the more alien a counterpart is, the greater the number of potential behavioural possibilities (both positive and negative). Aliens are so alien to us that their behaviour towards us cannot be anticipated in any way. For mankind they could be saviours, redeemers, bringers of salvation, but also merciless conquerors, cold-blooded destroyers and merciless rulers. Thus, from the outset, the figure of the alien in fictional-artistic works creates a tension between curiosity, hope and longing on the one hand, and fear, panic and despair on the other, which has been imaginatively and effectively introduced and realized in countless science fiction narratives. In other words:

> The strange and unknown arouses our curiosity, the strange and uncanny generates fear, the strange and promising is able to satisfy our longings. It is in this tension between different connotations and functions of the alien that the motif of the alien unfolds as a central element of science fiction in literature, film and television. (Hurst 2008, p. 33)

The concrete depictions of aliens and their relationship to humans in science fiction are correspondingly multifaceted: thus, throughout the history of the science fiction genre, we encounter aliens that are both vicious, conniving, and murderous as well as sympathetic, benevolent, and helpful. The fictional extraterrestrials confront humanity with.

> [...] positive as well as negative counter-designs to human nature and civilization. Peace-loving visitors from outer space, who fill humans with hope for a better future, exist in the fantastic film and TV worlds alongside aggressive creatures from distant star systems that cover our planet with war and destruction [...]. (Hurst 2008, p. 34)

The ratio between benign and malignant aliens in science fiction is anything but balanced: For every one alien species with good intentions, there are approximately nine that are hostile to humans (Hurst 2004, p. 98). According to Engelbrecht (2008, p. 25), the various relationships between humans and aliens in science fiction can be mapped in a kind of 'phase space', which takes three dimensions into account: (1) Power: Do the aliens have more or less (especially technological) power in relation to humanity? (2) Strangeness of the aliens: expressed in terms of spatial distance from humans—the more similar the aliens are to humans, the closer they move to humans in the diagram. (3) Their behaviour: Do the aliens have good or bad intentions towards humans?

For each constellation within this phase space, there are countless examples from the now gigantic fund of stories about aliens (and humans). Consider, for example, the friendly, lovable, gnome-like alien in Steven Spielberg's 1982 science fiction tale *E.T.—The Extra-Terrestrial* on the one hand, and the nightmarish, insect-like extraterrestrial beings in the *Alien* film series on the other. In the 1994 science fiction film *Stargate,* an alien rules over human slaves like an Egyptian god, while the aliens in *Starship Troopers* (1997) wage war against humans as aggressive, insidious giant insects. In the universe of the film and television series *Star Trek* alone, we encounter hundreds of different alien species, ranging from life forms made of gas or merely energy, to cat-, bird- or reptile-like aliens, to trans-dimensional beings and the infamous 'Borg', a sinister, technologically advanced, half-organic, half-cybernetic species that possesses a collective consciousness and obliterates any form of individuality. The fictional universe of the *Star Wars* series, one of the most commercially successful film projects of all time, also features numerous different alien races engaged in an ongoing struggle between good and evil, embedded in the plot of a classic heroic epic (Fig. 2.3).

2.4 In the Hall of Mirrors

It would go far beyond the scope of this presentation to go into further details of the countless variants of extraterrestrial beings in science fiction. Rather, general tendencies and mechanisms with regard to the conceptions of fictional aliens are of concern here. In this context, it is first necessary to point out the paradoxical character that is fundamentally inherent in the artistic-literary design of aliens: the construction of the alien can only take place with recourse to the familiar, the own, the known. In other words:

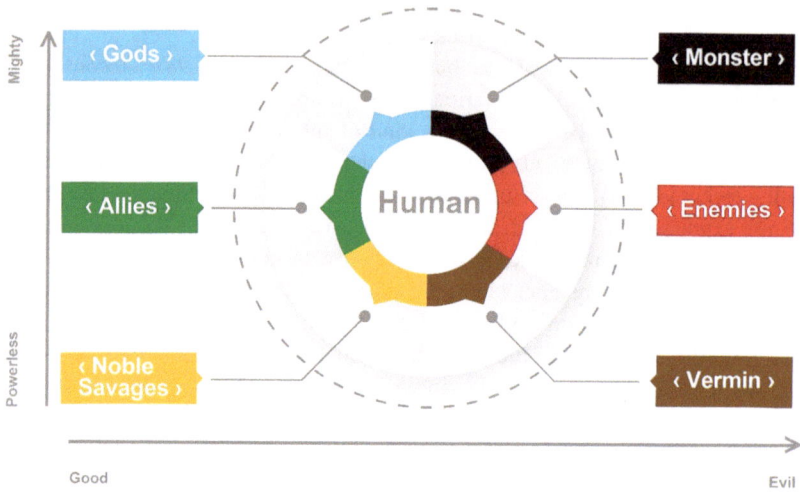

Fig. 2.3 Phase space of possible human-alien relationships in science fiction. *Source* Engelbrecht 2008, p. 25; redesigned by the authors for this volume

> The alien, the alien, becomes the touchstone of human imagination and the test of the limits of literary or cinematic representational power. How can one imagine, with the human mind caught up in its specific dimensions, something that is precisely not human and is supposed to exist beyond the given dimensions? How can one make the completely alien, the unimaginable, imaginable at all? Herein, of course, lies a fundamental paradox, for once the strange and unimaginable is made imaginable, it is no longer strange. (Hurst 2004, p. 98)

Thus, the fictional aliens do not reveal anything alien, new, unknown, but (inevitably) our own anthropological, psychological, cultural, political and social realities (Engelbrecht 2008, p. 14). The aliens in US science fiction films from the 1950s[8] are usually scary, destructive invaders with advanced weaponry. This reflects the collective fears of the US population of a Russian invasion and nuclear war (Hurst 2004, pp. 98–99). On closer inspection, the alien peoples in *Star Trek* are distorted images of US society:

---

[8] E.g. *The Thing from Another World* (1951), *Invaders from Mars* (1953), *The War of the Worlds* (1953), *Invasion of the Body Snatchers* (1956) and *The Blob* (1958).

Their understanding of politics, diplomacy, democracy and lifestyle in general becomes a grid to which (almost) all forms of extraterrestrial life have to submit; the fact that this is not always tolerant and free of prejudice, but that stereotypical thinking and one-sided, chauvinistic ideas are often transported with general concepts, is one of the rather alarming features of the 'Star Trek' universe. (Hurst 2004, p. 48)

The individual parts of the Alien series can be understood as confrontations with various socio-psychological sources of fear of their respective time of origin: from the arms race during the Cold War and biological weapons systems in the 1970 and 1980s to AIDS, genetic engineering and cloning experiments in the 1980 and 1990s to current debates about the dangers of artificial intelligence (Schmitt 2017, p. 37). In the highly successful TV series, *The X-Files* aliens form the starting point for the gigantic conspiracy of a small power elite and thus become a symbol for diffuse collective fears of political conspiracies. Even in narratives in which the encounter between humans and aliens is portrayed positively,[9] the aliens ultimately turn out to be a reflection of ourselves, as Hurst points out, because here, too, we can:

recognize forms of collective as well as individual projections; for not only the war-mongering and murderous aliens, but also the peacemaking travelers from other planets function in the feature films as projection surfaces of human affects and longings, they reflect, as in a fantastic conundrum, the human condition under specific historical, cultural, and social conditions. (Hurst 2004, p. 100)

Even in this brief overview, it is noticeable that aliens in science fiction are predominantly ascribed negative characteristics, which, in view of the projection processes explained, casts a questionable light on their creators. The all too often aggressive, destructive, insidious fictional aliens ultimately merely embody our own negative behavioral potentials. This in turn provides another explanatory dimension for the ambivalent basic relationship between humans and aliens in artistic-fictional representations:

The changing cultural images of extraterrestrials reflect more than anything else the continuing ambivalence of fascination and fear that the confrontation with the maximally alien can trigger in us. Fear, if we follow the psychological train of thought, results not only from fear of the unknown, but at least as much from fear of the all too familiar: of ourselves. It is ultimately the fear of what we see in the mirror. (Schetsche 2004, p. 18)

---

[9] Such as in *The Day the Earth Stood Still* (1951), *Close Encounters of the Third Kind* (1977), *Cocoon* (1985), *the Abyss* (1989) or *Arrival* (2016).

Our remarks should have made it clear that both (natural) philosophical and artistic-literary reflections on extraterrestrials are inevitably always determined by anthropocentric presuppositions, the respective state of knowledge, our purely earthly experiences, by social, cultural and political contexts, and probably also by human emotions. Thus, our conjectures about extraterrestrials can ultimately lead us nowhere, except always back to ourselves. But this is in no way to delegitimize creative thinking about 'the others', quite the contrary. The guiding principle can apply: 'Until the aliens are found, they must be invented' (Fetscher and Stockhammer 1997, p. 267). Our musings on the question of whether there are alien intelligences 'out there' form an almost inexhaustible source of immensely inspiring, stimulating, exciting narratives and, not least, have also triggered scientific and technical innovations—but one thing it has not yet provided us with is: Insights into potential *real* aliens. At present, we can say very little about them, as Hövelmann summarises:

> What would we be justified in assuming about an extraterrestrial, about whose real existence nothing at all needs to be asserted at this point and about which we would not possess the slightest empirical knowledge? Answer: nothing; at least nothing that would not already follow analytically from the designation 'extraterrestrial' itself - namely, that it is certainly a non-human being that just as certainly does not come from Earth. (Hövelmann 2009, p. 179)

Moreover, Hövelmann continues, "we can also make no educated guesses whatsoever about any technology developed by such a life form, the science it maintains, and its willingness and ability to engage in cosmic cooperation or opposition" (Hövelmann 2009, p. 179). Further, while it seems somewhat plausible to assume that possible intelligent extraterrestrials have radically different modes of appearance, life, and perception from us, ultimately even this remains conjecture. For the time being, when we think about extraterrestrials, we are in a kind of hall of mirrors—we usually see nothing but distorted images of ourselves (Anton and Schetsche 2014, p. 145). As long as we have no empirical knowledge about extraterrestrial civilizations, this will remain the case. This volume, however, would like to attempt to break out of this hall of mirrors as best we can, in order to explore what can be said—about potential extraterrestrial civilizations from a sociological perspective with at least sufficient justification—and what cannot be said.

References

Akerma, Karim. 2002. *Außerirdische Einleitung in die Philosophie. Extraterrestrier im Denken von Epikur bis Hans Jonas.* Münster: Monsenstein und Vannerdat.

Anton, Andreas & Michael Schetsche. 2014. Im Spiegelkabinett. Anthropozentrische Fallstricke beim Nachdenken über die Kommunikation mit Außerirdischen. In *Interspezies-Kommunikation. Voraussetzungen und Grenzen,* Hrsg. Michael Schetsche, 125–150. Berlin: Logos.

Bauman, Zygmund. 2000. Vereint in Verschiedenheit. In *Trennlinien. Imagination des Fremden und Konstruktion des Eigenen,* Hrsg. Josef Berghold, Elisabeth Menasse und Klaus Ottomeyer, 35–46. Klagenfurt: Drava.

Dick, Steven J. 1982. *Plurality of Worlds. The Origins of Extraterrestrial Life Debate from Democritus to Kant.* Cambridge: Cambridge University Press.

Engelbrecht, Martin. 2008. Von Aliens erzählen. In *Von Menschen und Außerirdischen. Transterrestrische Begegnungen im Spiegel der Kulturwissenschaft,* Hrsg. Michael Schetsche und Martin Engelbrecht, 13–29. Bielefeld: transcript.

Fetscher, Justus & Robert Stockhammer. 1997. Nachwort. In *Marsmenschen. Wie die Außerirdischen gesucht und erfunden wurden,* Hrsg. Justus Fetscher und Robert Sockhammer, 169–172. Leipzig: Reclam.

Gerritzen, Daniel. 2016. *Erstkontakt. Warum wir uns auf Außerirdische vorbereiten müssen.* Stuttgart: Kosmos

Günther, Ludwig. 1898. *Keplers Traum vom Mond.* Leipzig: Teubner.

Guthke, Karl S. 1983. *Der Mythos der Neuzeit. Das Thema der Mehrheit der Welten in der Literatur- und Geistesgeschichte von der kopernikanischen Wende bis zur Science Fiction.* Bern: Francke.

Heuser, Marie-Luise. 2008. Transterrestrik in der Renaissance: Nikolaus von Kues, Giordano Bruno, Johannes Kepler. In *Von Menschen und Außerirdischen. Transterrestrische Begegnungen im Spiegel der Kulturwissenschaft,* Hrsg. Michael Schetsche und Martin Engelbrecht, 55–79. Bielefeld: transcript.

Hoagland, Richard C. 1987. *The Monuments of Mars: A City on the Edge of Forever.* Berkeley: North Atlantic Books.

Hövelmann, Gerd. 2009. Mutmaßungen über Außerirdische. *Zeitschrift für Anomalistik* 9: 168-199.

Hurst, Matthias. 2004. Stimmen aus dem All – Rufe aus der Seele. Kommunikation mit Außerirdischen in narrativen Spielfilmen. In *Der maximal Fremde. Begegnungen mit dem Nichtmenschlichen und die Grenzen des Verstehens,* Hrsg. Michael Schetsche, 95–112. Würzburg: Ergon.

Hurst, Matthias. 2008. Dialektik des Aliens. Darstellungen und Interpretationen von Außerirdischen in Film und Fernsehen. In Von Menschen und Außerirdischen. Transterrestrische Begegnungen im Spiegel der Kulturwissenschaft, Hrsg. Michael Schetsche und Martin Engelbrecht, 31–53. Bielefeld: transcript.

Huygens, Christiaan. 1703. *Cosmotheoros oder Eine phantastisch-realistische Betrachtung der Schönheit der Welt, der Sterne und Planeten. Geschrieben von Christiaan Huygens für seinen Bruder Constantijn, Geheimrat der königlichen Majestät von Großbritannien.* http://www.passagenproject.com/christiaan-huygens-cosmotheoros.html.

Kaiser, Céline. 2004. „Fafagolik?" Fiktionen des Erstkontaktes in der ‚Marsliteratur' um 1900. In *Der maximal Fremde*, Hrsg. Michael Schetsche, 75–93. Würzburg: Ergon.

Kant, Immanuel. 1954. *Träume eines Geistersehers.* Berlin: Aufbau Verlag.

Kues, Nikolaus von. 1985. *De docta ignoratia.* Book II. http://wlym.com/archive/pdf/cusa_l earned02.pdf

Locke, John. 1999 [1690]. *An Essay Concerning Human Understanding.* The Pennsylvania State University. http://www.philotextes.info/spip/IMG/pdf/essay_concerning_human_u nderstanding.pdf .

Moore, Ben. 2014. *Da draußen. Leben auf unserem Planeten und anderswo.* Zürich: Kein & Aber.

Oeser, Erhard. 2009. *Die Suche nach der zweiten Erde. Illusion und Wirklichkeit der Weltraumforschung.* Darmstadt: WGB.

Plutarch. 1968. *Das Mondgesicht (De facie in orbe lunae).* Eingeleitet, übersetzt und erläutert von Herwig Görgemanns. Zürich: Artemis Verlag.

Schetsche, Michael. 2004. Der maximal Fremde – eine Hinführung. In *Der maximal Fremde. Begegnungen mit dem Nichtmenschlichen und die Grenzen des Verstehens*, Hrsg. Michael Schetsche, 13–21. Würzburg: Ergon.

Schetsche, Michael, René Gründer, Gerhard Mayer & Ina Schmied-Knittel. 2009. Der maximal Fremde. Überlegungen zu einer transhumanen Handlungstheorie. *Berliner Journal für Soziologie* 19 (3): 469–491.

Schmitt, Stefan. 2017. Das Wir da draußen. Wer nach Außerirdischen sucht, der findet – den Menschen. Im Kino, wo ein neuer ‚Alien-Film' anläuft. Aber auch in der Astronomie. *Die Zeit* Nr. 20 vom 11. Mai 2017: 37.

Wells, Herbert George. 1992. *The War of the Worlds.* https://gutenberg.org/files/36/36-h/36-h.htm.[Orig. 1898]

Wille, Holger. 2005. *Kant über Außerirdische. Zur Figur des Alien im vorkritischen und kritischen Werk.* Münster: Monsenstein & Vannerdat.

The Earth in Space   3

It is undoubtedly a sad fact that the beginning of space travel, the starting signal for the human 'reach for the stars', is located in the context of armed conflicts. The first human object to cross the boundary into space[1] was one of the A4/V2 rockets manufactured under the direction of Wernher von Braun at the Peenemünde Army Research Station in 1944. Although rocket development in the 'Third Reich' ultimately proved to be a bad investment from a military point of view due to the immense costs and the accuracy of the rockets, the Allies recognised the potential of the technology behind it and transferred rocket components, production facilities and scientists and technicians involved in rocket development to the Soviet Union and the USA. There, German expertise in rocket technology formed the basis for the military and space technology developments of the next decades.

In the confrontation between the USA and the Soviet Union that began immediately after the Second World War, rocket technology gained a propagandistic significance in addition to its military value in the context of space-flight possibilities. The competition between the two systems was supplemented by a 'race into space'; technical developments in space flight became a further measure of capability or technical superiority, and both sides endowed their respective space programmes with generous funds. Spaceflight thus became an "instrument of a race for world supremacy, conducted by symbolic means" (Weyer 1997, p. 465). As a result, there was a rapid technical development and the space programs of both competing systems were able to record successes and highlights for themselves in rapid succession: In October 1957, the Soviet Union launched *Sputnik*,

---

[1] In aeronautics and astronautics, space is usually referred to as from an altitude of 100 km above sea level (Kármán line). Above this altitude, the Earth's atmosphere is so thin that it can no longer be used for lift or propulsion.

© The Author(s), under exclusive license to Springer Fachmedien Wiesbaden Gmbh, part of Springer Nature 2023
A. Anton and M. Schetsche, *Meeting the Alien*,
https://doi.org/10.1007/978-3-658-41317-0_3

37

the first satellite; a month later, *Laika*, a dog, was the first living creature; and in 1961, Yuri Gagarin became the first human to orbit the Earth. In 1969 the USA achieved the first manned moon landing, in 1971 the Soviet Union placed the first manned space station *Salyut 1* in space. With the *Space Shuttle*, the USA created the first reusable space vehicle in 1981.

These successes in the field of space technology also spurred on thinking about extraterrestrial life, as they proved in practice that the limits of the Earth can be overcome by technical means. At least the near-Earth space became part of the human sphere of influence (Schetsche 2005)—and what mankind now succeeded in doing, an extraterrestrial civilization could have succeeded in doing before: leaving its own home planet and penetrating the vastness of space by technical means. Against this background, the questions that had preoccupied philosophers and natural scientists for centuries took on new weight: Does the Earth occupy a special position in the universe or does life arise everywhere where the basic conditions for it exist? How probable is the emergence of living beings that can act in a planned manner and consciously come to terms with their own existence? How far away from Earth might extraterrestrial civilizations be? Are they technically and culturally more advanced than humanity? Is there a chance of contact? And what would this mean for our human self-image? (Pirschl and Schetsche 2013, p. 30).

These questions were also on the minds of physicists Giuseppe Cocconi and Philip Morrison, who in 1959 published an article in the scientific journal *Nature* entitled *Searching for Interstellar Communications* (Cocconi and Morrison 1959, pp. 844–846). Even though, as the two physicists first stated at the beginning of the article, there was currently no way to estimate (1) the number of planets in the universe, (2) the probability of the emergence of life, and (3) the emergence and development of extraterrestrial civilizations, they did think it likely that there were intelligent extraterrestrial species that were technologically superior to us and, furthermore, interested in interstellar communication with humanity. Cocconi and Morrison therefore asked:

> We shall assume that long ago they established a channel of communication that would one day become known to us, and that they look forward patiently to the answering signals from the Sun which would make known to them that a new society bas entered the community of intelligence. What sort of a channel would it be? (Cocconi and Morrison 1959, p. 844)

They suggested using radio telescopes to search for messages from other technological civilizations in our galaxy. This, of course, raised the question of which frequency range to search. Cocconi and Morrison offered a solution to this as well: Alien messages, they suggested, could be transmitted on a radio frequency of 1420 MHz, which corresponds to a wavelength of 21 cm. This is the wavelength of the radio radiation of neutral hydrogen, the most common element in the universe, and should, Cocconi and Morrison continued, also appear to an extraterrestrial civilization as a suitable frequency:

> It is reasonable to expect that sensitive receivers for this frequency will be made at an early stage of the development of radio-astronomy. That would be the expectation of the operators of the assumed source, and the present state of terrestrial instruments indeed justifies the expectation. Therefore we think it most promising to search in the neighborhood of 1,420 Mc./s. (Cocconi and Morrison 1959, p. 845)

About two years later, the astrophysicist Frank Drake developed an equation that was to be used to estimate the number of technically developed extraterrestrial civilizations in our galaxy (Krauss 2002, p. 29). In the meantime, the so-called *Drake equation* has gained some notoriety, even outside the narrow circle of expert astronomers. The equation reads:

$$N = R * f_p * n_e * f_{el} * f * f_{ic} * L.$$

N stands for the possible number of intelligent extraterrestrial civilizations in our galaxy at the present time, R for the mean star formation rate, f_p for the proportion of stars with planetary systems, n_e for the number of planets in the so-called habitable (i.e. life-friendly) zone. The parameter f_l indicates the proportion of planets on which life exists, f_i the proportion of planets with intelligent life, and f_c is intended to capture the proportion of alien civilizations interested in extraterrestrial communication. Finally, factor L is used to account for the average lifespan of technologically developed alien civilizations (Shostak 1999, pp. 218–219).

The main problem with the Drake equation is that the probability values of some factors seem almost indeterminable. How, for example, should one arrive at a meaningful estimate with respect to planets with intelligent life or the average lifespan of extraterrestrial civilizations? Drake himself therefore referred to his equation as a "compounding of uncertainties" (Drake and Sobel 1992, p. 45). For this reason, the Drake equation should be viewed as food for thought rather than a tool for actually calculating the number of extraterrestrial civilizations. Accordingly, the number of currently communicating extraterrestrial civilizations in our galaxy estimated using the Drake equation varies from one (namely humanity itself) to several million. Drake himself estimated the number at 10,000 (Shostak 1999, p. 222)—which, however, was an almost arbitrary value given the state of knowledge at the time. But even today's estimates, despite all the advances in astronomy and astrobiology, are not much more certain because of the general problems described. Nevertheless, it should be emphasized that today, more than half a century after the equation was written, at least the first three factors of the equation can be determined with increasing accuracy. In order[2]:

3.1   The Star Formation Rate

The star formation rate (R) indicates how many stars are formed per year in a certain region of space (here: in the Milky Way). It depends on various factors such as the age of the galaxy, the interstellar mass within the galaxy and the local physical properties of the interstellar mass (e.g. density, temperature, magnetic field strength). Its value can be estimated relatively well by empirical observations and astrophysical models (see for example Kennicutt and Evans 2012). According to recent calculations by NASA and ESA, the current star formation rate in the Milky Way is about 0.7–1.5 solar masses per year, which is equivalent to 1.5–3 stars per year if the average mass of a new born star is 0.5 solar masses (Robitaille and Whitney 2010). However, the overall average star formation rate of the Milky Way (i.e., since its existence) is 10–20 stars per year. This shows that the number of newly forming stars is not constant, so galaxies are not always equally 'fertile' and sometimes produce more and sometimes fewer stars. It should be noted in this context that not all stars are suitable for producing potentially life-friendly conditions in their environment. Stars that have greater mass and luminosity than the Sun use up their energy too quickly, leaving insufficient time for complex life forms to evolve. Stars with lower mass and luminosity than our Sun, known as *red dwarfs,* which include about two-thirds of all stars in our galaxy, have long lifetimes, but this would require planets to orbit very closely around them to be in the habitable zone. This in turn creates two problems: First, such planets are exposed to high doses of hostile solar radiation, and second, strong gravity, which can lead to synchronous (or locked) rotation, so that one half of the planet is always facing its home star, and the other half is always facing away (like

---

[2] We present the astrophysical and astrobiological backgrounds of our sociological considerations in this chapter in such detail because they do not belong to the usual canon of social scientific knowledge. It should be added that the current state of scientific research provides a central justification for our decision to make such a strong case for the idea of an exosociology at *this point in time.*

the Moon facing Earth). Among astrobiologists, however, it is discussed whether dense atmospheres or large oceans on planets in the vicinity of red dwarfs could lead to a temperature equilibrium on the surface and thus to more favourable conditions for the emergence of life, which in the end does not rule them out per se as potential candidates for inhabited planets (Zaun 2006; Joshi 2003; Scalo et al. 2007).

Of particular interest, from an astrobiological point of view, are those stars that are similar in mass and burning time to our own Sun. Specifically, these are stars of spectral classes G and K.[3] From today's perspective, such stars are best suited to create a potentially life-friendly environment. Current estimates suggest that about 15% of all stars fall into these categories (Leibundgut 2011, p. 137). So, what does this mean for estimating a value for the variable R of the Drake equation? Since we do not know whether the emergence of life is actually tied to specific classes of stars, this estimate is extremely difficult and the value can vary greatly. If we assume that only G and K stars can actually be considered for the formation of life, this means that in our Milky Way a star is currently formed approximately every three to six years in whose environment life could form and exist. In relation to the total number of stars in the Milky Way (about 150 billion), this would result in a number of about 22 billion potentially life-friendly suns. At an average star formation rate of 15 stars per year, the value for R would be $15 \times 0.15 = \mathbf{2.25}$.

3.2 How Many Planets Are There in the Milky Way?

The next factor in the Drake equation f_p asks about the fraction of stars that are orbited by a planetary system. For a long time, it was not clear how often planetary systems occur in space and whether they are the exception rather than the rule. The first definitive evidence of an *exoplanet*, i.e. a planet outside our solar system, was found in 1995 by the Swiss astronomers Michael Mayor and Didier Queloz. The planet is called *Dimidium* (or 51 Pegasi b) and orbits the star 51 Pegasi, about 40 light-years away in the constellation Pegasus (Mayor and Queloz 1995). Since then, the events have virtually overtaken each other: Only one year after the discovery of *Dimidium*, more exoplanets were known than there

---

[3] The spectral class (also spectral type) is a classification system dating back to nineteenth century astrophysics, which classifies stars according to their light spectrum. The vast majority of stars (over 90%) belong to the seven spectral classes (so-called basic classes) O, B, A, F, G, K and M, which at the same time represent a temperature sequence from high (B) to low temperatures (M).

Fig. 3.1 Exoplanets (artist's rendering): The diversity of alien worlds. *Source* original graphic for this volume by Nadine Heintz

are planets in our solar system. After Mayor's and Queloz's discovery (which he confirmed by his own measurements), the US astronomer and astrophysicist Geoff Marcy predicted that from now on, with continued observation, new planets would 'just hail down'. He was to be proved right (Shostak 1999, p. 81). In the early days of exoplanet detection, an average of about one planet was discovered per month, and now new planets are being added almost daily. In 2016 alone, 1,464 new exoplanets were registered.[4] Today (as of March 2018), around 3,700 planets are known to exist outside our solar system (Fig. 3.1).

The increased number of discoveries is mainly due to improvements and innovations in detection methods. With the so-called *transit method*, for example, minimal obscurations are measured that occur when a planet passes its home star on the line of sight to Earth. The amount of starlight blocked by the planet is proportional to the size of the obscuring planet (see for example Moore 2014, p. 35). In 1999, the first exoplanet was discovered using the transit method. It bears the designation *HD 209,458 b* and orbits a star in the constellation Pegasus at a distance of about 150 light-years from Earth. In the meantime, the transit method has become the most proven means of detecting exoplanets. It has been used to discover about 80% of all known planets outside our solar system.[5] It must be

---

[4] Own calculation based on data from http://www.exoplanet.eu/.

[5] See NASA Exoplanet and Candidate Statistics: https://exoplanetarchive.ipac.caltech.edu/docs/counts_detail.html.

taken into account that the probability of observing a planetary transit at a randomly selected star is less than one percent, since the planet's orbit is most likely such that its passage past its home star cannot be observed from Earth. The fact that so many exoplanets could nevertheless be registered using the transit method suggests that planetary systems are common. Today, it is assumed that on average every star in the Milky Way has at least one or two planets, which would mean that there are more than a hundred billion planets in our cosmic environment alone (Cassan et al., 2012; see also Wandel 2014; Scholz 2014). Assuming an average of one planet per star, the corresponding value for the factor f_p in the Drake equation would be 1. We can therefore state: $f_p = 1$.

3.3 Planets in the Habitable Zone

In astrobiology, liquid water is considered one of the most important, if not indispensable (see for example Janjic 2017, p. 207; Benner et al. 2004), building blocks of life. For liquid water to exist on a planet, it must be at a suitable distance from its home sun: If it is too close to the star, the water evaporates completely; if it is too far away, it freezes permanently. In our solar system, the habitable zone (also known as the *ecosphere*) ranges from about 120–220 m km from the Sun. Earth, at a distance of 150 m km from the Sun, is roughly in the middle of the habitable zone, Venus, at a distance of 108 m km from the Sun, is in front of it, and Mars, at a distance of 228 m km from the Sun, is just behind it (Leibundgut 2011, p. 138).

The first known exoplanets for which it has been discussed whether they lie in the habitable zone of their home planets are the aforementioned planet *HD 209,458 b* and the planet *Gliese 581 c,* discovered in 2007. *HD 209,458 b* is a so-called 'hot Jupiter', i.e. a gaseous planet that is assumed to have a fairly high surface temperature due to a close orbit around its home sun. In 2007, large amounts of water vapour were detected in the atmosphere of *HD 209,458 b,* making it the first exoplanet where water has been detected. In later investigations, methane and carbon dioxide were also registered in the planet's atmosphere. However, the existence of life on *HD 209,458 b is* considered impossible (Kayser 2009). *Gliese 581 c* orbits a red dwarf in the constellation Libra and has about five times the mass of Earth. So far, no water has been detected on the planet. Furthermore, life on *Gliese 581 c is* considered almost impossible, since its home star *Gliese 581* emits high doses of X-rays at irregular intervals (Selsis et al. 2008).

Kepler-22b, the first exoplanet on which life could theoretically exist, was reported in 2011. It is about 2.4 times larger than Earth and about 600 light years away from Earth. However, it is not yet clear whether *Kepler-22b is* a gas planet or a rocky planet. Also in the habitable zone is the planet *Kepler-452b*, discovered in 2015, 1,400 light-years away in the constellation Swan, which orbits a Sun-like star, presumably has a solid surface and—at least theoretically—could be Earth-like (Gerritzen 2016, p. 99). In total, 53 exoplanets have been discovered so far within the habitable zone of their central stars, assuming an optimistic interpretation of the available data—although these are very likely to include some gaseous planets.[6] On the basis of various data (such as radius, density, surface temperature) the so-called *Earth Similarity Index* (ESI) is calculated for these planets, which represents a rough measure for the evaluation of the potential Earth similarity of exoplanets. ESI values range from 0 to 1, with Earth itself having a value of 1. The 'hottest candidate' so far for a high Earth similarity and thus for the existence of life is the planet *Proxima Centauri b*, with an ESI value of 0.87 (Janjic 2017, p. 10). Curiously, the planet is located—at least by cosmic standards—in our immediate neighbourhood: it orbits the *star Proxima Centauri* (or Alpha Centauri C), which together with the stars *Alpha Centauri A* and *B* forms a triple star system and is the closest star to the Sun at a distance of 4.2 light-years. *Proxima Centauri b* has about 1.27 times Earth's mass and liquid water may exist on its surface (see Boutle et al. 2017). Another interesting exoplanet is *Kepler-442b*, a rocky planet orbiting its home star *Kepler-442* in the constellation Lyra in its habitable zone, has about 1.3 Earth radii and could also host liquid water.

The crucial question is how often the orbit of exoplanets lies in the habitable zone of their home stars. Conservative estimates assume that this is the case for about 1–10% of all exoplanets (Leibundgut 2011, p. 138). Statistical analyses based on data from the Kepler mission concluded that there could be as many as 40 billion Earth-like planets within the Milky Way in the habitable zones of Sun-like stars and red dwarfs, including 11 billion orbiting a star similar to our Sun (Petigura et al. 2013). In addition, it must be taken into account that life could in principle also be possible *outside the* habitable zone. Within our solar system, the Jupiter moon *Europa, for* example, is a possible candidate. *Europa* is completely covered by water ice, is located far outside the habitable zone, and its surface has an average temperature of up to minus 170 °C. However, due to Jupiter's massive gravitational forces, there could be an ocean of liquid water

---

[6] See University of Puerto Rico at Arecibo: Habitable Exoplanets Catalog: http://phl.upr.edu/projects/habitable-exoplanets-catalog.

beneath its icy surface—and perhaps life within it (Moore 2014, pp. 196–204). Other candidates for life within our solar system are Saturn's moons *Titan* and *Enceladus*. This means that life could in principle also have arisen in more remote regions of planetary systems.

What value for the factor n_e of the Drake equation can we now determine on the basis of these considerations? In view of the statistical analyses based on the Kepler data, planets in the habitable zone of their central star seem to be quite common. It does not seem unrealistic to assume that a life-friendly planet (in the habitable zone or outside of it) could exist in every other planetary system. However, we again want to estimate conservatively and set the value much lower. We assume that on average there is a planet with basically life-friendly conditions in every twentieth planetary system. This means: $n_e = 0.05$.

3.4   The Origin of Life

Given our current knowledge and understanding, the estimation for this factor of the Drake equation becomes highly difficult and speculative. Nevertheless, we want to try it. The next factor f_l indicates proportionally those potentially life-friendly planets on which life has actually arisen. To date, there is no conclusive scientific explanation for the origin of life on Earth. In particular, the leap from inorganic chemical compounds to organic biological life continues to puzzle science (Gerritzen 2016, p. 100). Even the definition of life presents certain difficulties, since various criteria for life also apply to inanimate systems such as crystals, technical systems, computer programs (including computer viruses), or fire. Nevertheless, there is a broad consensus in biology with regard to the basic physico-chemical properties that a living system must have (Koshland 2002):

(1) Homeostasis: regulatory mechanisms to maintain states of equilibrium;
(2) Cells as the basic unit;
(3) A metabolic system (metabolism);
(4) Growth and thus the ability to develop;
(5) Adaptability to changing environmental conditions (through evolutionary processes);
(6) Stimulus openness or sensory: the ability to respond to chemical or physical stimuli;

(6) Reproduction and genetic variability.

The problem with this catalogue of criteria is that it excludes hypothetical early or preliminary forms of life as well as borderline cases such as viruses. In the search for life in space, however, pre- and borderline forms of life are also of great interest, which is why there are proposals for more open definitions of life on the part of astrobiology. For example, Schulze-Makuch and Irwin define life as follows:

> life is (1) composed of bounded microenvironments in thermodynamic disequilibrium with their external environment, (2) capable of transforming energy and the environment to maintain a low-entropy state, and (3) capable of encoding information. (Schulze-Makuch and Irwin 2004, p. 14)

Astrophysicist Moore (2014, p. 52) defines life simply as "Any molecular structure capable of carrying within itself the information and mechanism necessary for reproduction."

The next question is what conditions must be met for life to arise. There are various approaches to this in biology or astrobiology, but fundamentally the following factors are considered to be absolutely necessary for the emergence of life: (1) the presence of a solvent for nutrients (e.g. water); (2) energy (e.g. in the form of light, hydrothermal vents or lightning); (3) various chemical substances such as carbon, hydrogen, nitrogen, phosphorus, oxygen, sulphur, iron, etc.; and (4) suitable temperatures so that a liquid solvent is available at least temporarily (Leibundgut 2011, pp. 139–140).

The earth is about 4.5 billion years old. The history of the earth is divided into so-called *eons*. The formation of the earth up to 4 billion years ago is called the *Hadean*, up to 2.5 billion years ago the *Archean*. In these early phases of the earth's history extreme, one could think, hostile conditions prevailed on our planet. The earliest traces of life on Earth are 3.8 billion years old. Life thus evolved at an astonishingly early age—as early as 700 million years after the Earth came into being (Gerritzen 2016, p. 100). This justifies the frequently formulated assumption that life arises *automatically*, as it were, as soon as suitable conditions are present. Occasionally, this postulate is referred to as *biological determinism* (Davies 1999, p. 11). The problem is that the fact that life has arisen on Earth hardly allows any conclusions to be drawn about the probability of life arising outside Earth. The discovery of even a single alien life form would fundamentally change this: Evidence that favourable environmental conditions twice independently led to the emergence of life would be a strong argument for biological determinism. Clues to clarify this question could perhaps be found on Earth as well: It would be entirely possible that life on Earth arose not once, but several times, and is currently still arising. Thus, it is quite conceivable that, in the sense of *continuous abiogenesis,* protocellular units develop again and again, but then cannot develop into functional cells because they are previously consumed and used up by microorganisms (Janjic 2017, p. 208). Evidence that life has already formed on Earth several times would support the hypothesis that life arises with a high probability under suitable conditions. So far, however, no clear evidence for this has been found (Davies 2007).

It has been clear for some time, however, that the formation of organic chemical compounds in space, which in turn are an important basis for the emergence of life, occurs frequently and that there is even a *tendency* towards the development of complex organic molecules, as the German astronomer Dieter B. Herrmann explains:

> Above all, it is evident that there is a direction of preference in the evolution of chemical compounds in the universe which leads from the simplest molecules to more complicatedly constructed organic molecules, and which prevails with such power that even apparently unfavorable external conditions cannot change it. In such cases one speaks of preferences, which play an important role in physics and chemistry. Preferences lead to the fact that in gas mixtures not every arbitrary combination of molecules unites to steady compounds, but that *selected* reactions occur with decidedly higher probability. It is thus a matter of a favoring of certain chemical processes over all others, arising on the basis of elemental properties. (Herrmann 1988, p. 165; emphasis in original)

Such chemical processes can take place not only on planets, but also on asteroids and in clouds of gas and dust in interstellar space. All in all, organic chemical compounds in space prove to be almost "mass-produced goods of the cosmos" (Herrmann 1988, p. 164). Against this background it is discussed whether life on Earth had its beginning here at all or was 'imported' from outside, as it were. Ben Moore, for example, considers it quite probable,

> [...] that the most suitable environment in which primitive life could have evolved was not on Earth. I prefer the idea that life got its start on a giant asteroid or a dwarf planet. And that such an object collided with Earth at a point in time and 'delivered' life here when conditions allowed it to flourish. (Moore 2014, p. 132)

Ideas of this kind are called the *panspermia hypothesis.* The idea of panspermia stands or falls on the question of whether simple life forms or precursors of life

could survive for long periods of time under space conditions and survive a colli-
sion of their carrier object with a planet. To answer this question, it helps to take
a look at life forms on Earth that live under extreme conditions and are therefore
called *extremophiles in* science. Here we see something astonishing: life on Earth
has conquered almost every corner of the planet, even those that at first glance
seem exceedingly hostile to life. For example, microbes capable of surviving
temperatures of up to plus 300 °C have been discovered near hydrothermal vents
on the ocean floor. Organisms that do not need sunlight to survive live in caves,
deep in the ground and at the bottom of the oceans. The bacterium *Planococcus
halocryophilus has a* kind of natural 'frost protection' and survives temperatures
down to minus 37 °C, while other bacterial species are able to survive high doses
of ionized radiation (Moore 2014, pp. 147–160).

The cute-looking tardigrades (Tardigrada), which are about one millimetre in
size, prove to be true 'survival artists'. They are found almost everywhere on
earth and seem to be almost 'indestructible'. When their environment becomes
too dry, the animals go into a kind of suspended animation (cryptobiosis) in
which their metabolic processes are reduced to an absolute minimum. In very cold
temperatures, they may even freeze temporarily (cryobiosis) and their metabolism
comes to a complete halt. As soon as temperatures rise, the tardigrades 'wake
up' again and go into an active state. In experiments, tardigrades were frozen for
20 h at temperatures of minus 272.95 °C (this almost corresponds to the lower
limit of the temperature scale), they were stored for 20 months at minus 200 °C,
but also at high temperatures up to plus 150 °C. They were exposed to extreme
pressures and to toxic gases, including carbon monoxide, carbon dioxide, nitrogen
and sulphur dioxide. The results of the experiments were always the same: After
the tardigrades were returned to a more life-friendly environment, they awoke
and continued to live as if nothing had happened. Such resilient organisms as
the tardigrades might well be capable of surviving a journey through space on
an asteroid and its impact on another planet (Moore 2014, pp. 160–163)—this
also applies to various species of bacteria. What is certain today is that organic
compounds such as amino acids can survive the extremely hard 'landing' on Earth
on a meteorite. For example, a large number of organic compounds were found in
material from the meteorite *Murchison,* which fell in Australia in 1969, including
so-called *diaminocarboxylic acids*, which are considered to be the building blocks
of the first genetic material on Earth (Meierhenrich et al. 2004) (Fig. 3.2).

In summary, it can be stated that with regard to the current state of knowledge
about the persistence of terrestrial life forms and the transferability of organic
compounds between different celestial bodies, the panspermia hypothesis cannot
be dismissed out of hand, even if it is quite rightly pointed out again and again

Fig. 3.2 True survivors: a tardigrade. *Source* public domain image[7]

that this is so far pure speculation (Janjic 2017, p. 85). What is clear is that the possibility of panspermia would mean that life on other planets does not necessarily have to have originated there. This raises an interesting question to be answered on the part of astrobiology: Could life, assuming transmission by panspermia, also spread to places in the universe where it could not have arisen because of unfavorable conditions?

Returning to the Drake equation, an estimate of the probability of the origin of life on potentially life-friendly planets contains many imponderables and is therefore extremely difficult. However, three factors lead us to believe that the origin and maintenance of life is more likely than unlikely under suitable conditions: (1) the early origin of life on Earth; (2) the frequency with which basic building blocks of life, such as organic compounds, occur in the universe; and finally, (3) the general resilience and adaptability of life. This leads us to assume that at least half of all life-friendly planets actually harbor life. This would make $f_l = 0.5$.

3.5 How Does Mind Come into the World?

At least as difficult to answer is the question of how likely it is that *intelligent life* will evolve from life. Regardless of the fact that even a definition of intelligence is anything but trivial, it should be clear that the Drake equation is aimed at

---

[7] Available online at: https://de.wikipedia.org/wiki/B%C3%A4rtierchen.

those forms of potential extraterrestrial life forms that have developed a *civilization or technological intelligence*. According to astrobiologists Schulze-Makuch and Bains, there are four essential characteristics of a technological intelligence: (1) a sufficiently complex neural structure for intelligence to develop, (2) the ability to manipulate the environment (using prehensile organs and tools), (3) the ability to use and control environmental or energy resources (e.g., through fire), and finally (4) the ability to interact socially (exchange ideas, etc.) and to cooperate systematically (e.g., in the form of trade) (Schulze-Makuch and Brains 2017, p. 170).[8] If one follows this conception, the following distinction can be made: There are undoubtedly a multitude of intelligent life forms on Earth, but only one of them—humans—has so far developed technological intelligence or a civilization. While anthropological research paints an increasingly nuanced picture of why the species *Homo sapiens*, a mammal from the order of *primates*, the suborder of *dry-nosed primates*, and the family of *great apes*, of all species, evolved the capacity for technological intelligence, it is entirely unclear how *likely* this evolutionary process was. Here, opinions vary widely. In essence, the relevant controversies revolve around the question of whether the mental abilities of humans are the result of more or less *random* and thus rather improbable developmental processes—one could also say: a random by-product of evolution—or whether conscious intelligence or 'mind' is an inherent and thus, under certain conditions, a probable developmental potential of evolutionary processes. For example, the evolutionary biologist Ernst Mayr assumes that the emergence of life—and in particular of intelligent life forms—is highly improbable and points out that of the approximately 50 billion species that evolution has so far produced on earth, only *one* species has developed a form of intelligence that has made a complex form of civilization possible:

> How many species have existed since the origin of life? This figure is as much a matter of speculation as the number of planets in our galaxy. But if there are 30 million living species, and if the average life expectancy of a species is about 100,000 years, then one can postulate that there have been billions, perhaps as many as 50 billion species since the origin of life. Only one of these achieved the kind of intelligence needed to establish a civilization. (Mayr 1995)

Proponents of the thesis that the emergence of life and intelligence is likely under favourable conditions refer, among other things, to the principles of *averageness* and *uniformitarianism* (Anton and Schetsche 2015, p. 28). According to the principle of averageness, the evolution of life and intelligence as it has occurred on

---

[8] We will revisit and discuss this conception in more detail in Chap. 10.

Earth should not be understood as something unique. Rather, it is assumed that the evolution of life and intelligence actually takes place wherever suitable environmental conditions exist. In addition, the principle of uniformitarianism claims that the same laws of nature apply everywhere in the universe—this would not only lead to an identical structure of the cosmos (e.g. with regard to the relationship between matter and energy), but would also make the development of life and intelligence on Earth the "result of a natural development of physical processes in the cosmos" (Heidmann 1994, p. 131). Consequently, this would mean that the development of life and intelligence at suitable places in the universe could have proceeded in a very similar way as on Earth—or even *must* have proceeded (von Hoerner 2003, p. 12 and p. 55; Sheridan 2009, p. 24).

In our opinion, a highly interesting impulse with regard to the discussion about the question of the probability of the emergence of intelligence and consciousness came a few years ago from the US-American philosopher Thomas Nagel. In his book *Mind and Cosmos*, Nagel argues that the currently dominant materialistic understanding of nature is not capable of satisfactorily explaining the emergence of consciousness. Specifically, (1) mental states could not be attributed entirely to physical–chemical properties in the brain within the framework of psychophysical reductionism favored by the natural sciences, and (2) evolutionary theory could not adequately explain the genesis of consciousness within the history of human development. But since the existence of consciousness is one of the fundamental realities of nature, the previous understanding of nature is wrong or at least incomplete. In Nagel's words:

> The existence of consciousness is both one of the most familiar and one of the most astonishing things about the world. No conception of the natural order that does not reveal it as something to be expected can aspire even to the outline of completeness. (Nagel 2014, p. 53)

An approach that adequately accounts for the phenomenon of consciousness, Nagel argues, must include both a *constitutive account*, showing the ways in which complex material systems (such as the human brain) are also mental, and a *historical account*, explaining how and why the mind came into the world in the first place. For this, in very rough summary, there are two possible answers: an *emergence-theoretic* one and a *reductive* one. The emergence-theoretic approach says, simply put, that consciousness is an emergent property of the brain, that is, a property that *arises* from the structure of the human brain without being fully traceable to the properties of its individual elements. In other words, the whole is more than the sum of its parts! The weak point of this approach is that the moment of emergence remains fundamentally unexplained, as Nagel notes:

> That such purely physical elements, when combined in a certain way, should necessarily produce a state of whole that is not constituted out of the properties and relations of the physical parts still seems like magic even if the higher-order psychophysical dependencies are quite systematic. (Nagel 2012, pp. 55–56)

Nagel therefore favours a reductive (not to be confused with reductionist) approach, according to which the basic properties of a system must in principle be able to be explained from the properties of its individual components. With respect to consciousness, this has far-reaching consequences, for it means that the basic elements of our brain, i.e. atoms and molecules, must have *mental properties* that are capable of explaining the nature of human consciousness. Such ideas are called *panpsychism* and are explicitly brought into play by Nagel:

> Everything, living or not, is constituted from elements having a nature that is both physical and nonphysical – that is, capable of combining into mental wholes. So this reductive account can also be described as a form of panpsychism: all the elements of the physical world are also mental. (Nagel 2012, p. 57)

Nagel speaks in this context of 'protomental' properties of matter, which would carry the emergence of spirit or consciousness as a potential possibility of development, just as atoms and molecules are capable of forming galaxies on a higher level.

No doubt this approach is highly speculative and presuppositional, as Nagel himself acknowledges, yet in our view it exhibits a high degree of plausibility and logical rigor. The implications of this model are enormous: what we call intelligence, mind, or consciousness would not be an accidental or by-product of evolution, but an *inherent property of the universe* that appears wherever sufficiently complex structures (be they natural or artificial) allow it to unfold. This model is also captivating because it is able to incorporate and even predict different forms of consciousness or intelligence (e.g. in animals). With regard to the question of extraterrestrial intelligences, the position could be derived from this perspective that everywhere in the universe, where more complex forms of life have arisen, intelligence or processes of consciousness can *automatically* be found. We would like to call this point of view, in reference to the idea of biological determinism, *ratiogenetic determinism*.[9]

---

[9] Such a position is advocated by Martinez (2014), for example.

We explicitly *disagree with* this view, as we ultimately consider it too presuppositional and, at its core, anthropocentric. Instead, we assume that the emergence of intelligence and consciousness is not an *inevitable goal of* evolutionary processes, but rather a *potential for development*, which, however, is realized with a relatively high probability, also due to evolutionary advantages. We therefore estimate, to return to the Drake equation, that on half of all planets that have given rise to life, after an appropriate evolutionary period (more on this later), intelligent, self-aware life has also arisen. This means: $f_i = 0.5$.

3.6 The Communication Readiness of the Extraterrestrials

How many of these intelligent alien civilizations will wonder, as we do, if they are alone in space and have an interest in interstellar communication? Again, in the end, we can only speculate. This factor of the Drake equation is a little misleading, however, because Drake was concerned here not only with extraterrestrial civilizations that have an active interest in interstellar communication, but also with those that might be detected by us on the basis of their technical signatures: "Even if they weren't thinking of communication with galactic neighbours, their capacity for communication might give them away." (Drake and Sobel 1992, p. 60). Thus, the factor f_c basically describes the proportion of extraterrestrial civilizations that, due to their technological development, have left traces that can be detected by us *and* that may also have an interest in interstellar communication. In this context, the considerations of the Russian astronomer Nikolai Kardashov are interesting, who in the mid-1960s proposed a categorization of the development stages of extraterrestrial civilizations according to their energy consumption, the so-called *Kardashov scale*. The basic form of the scale envisages three types of extraterrestrial civilizations:

- Type 1: Civilizations that are capable of using all the energy available on a planet. On Earth, this corresponds to about 1.74×10^{17} W.
- Type 2: Civilizations that are capable of using the total power of their home star. This corresponds to approx. 4×10^{26} W.

- Type 3: Civilizations that are able to use the energy of an entire galaxy. This is approximately 4×10^{37} W (Gerritzen 2016, pp. 75–76).

An extraterrestrial civilization with a high energy utilization capability could theoretically send out strong electromagnetic signals to establish interstellar contact—provided it *wants to* do so in the first place. Beyond that, however, we could of course also receive signals that were not sent with the intention of establishing communication, but in connection with the application of certain technologies. What might such an extraterrestrial technosignature look like?

For some years the unusual star *KIC 8,462,852*, also known as *Tabby's Star*, has caused some excitement. This star is unusual because it exhibits strange dimming phases that have never been observed before in any other star. The star temporarily loses up to 20% of its total brightness, only to return to 100%. In addition, analyses of older images of the star seemed to indicate that the dimming of the star *increases* over the years. There are several explanations for Tabby's Star's mysterious dimming, ranging from comets or even planetary collisions with the star that left dust clouds behind, to the theory that the star is on the border between two different physical states and therefore exhibits fluctuations in brightness. Most interesting in our context, however, are speculations that the fluctuations are due to a so-called alien *megastructure*, i.e. an artificial structure built by extraterrestrials. The latter hypothesis is based on the idea that extraterrestrial civilizations, once they have reached a certain level of technological development, could switch to intercepting their energy directly from their home star and install a gigantic megastructure, also called a *Dyson sphere, in* its vicinity for this purpose. Thus, in terms of the Kardashov scale, one would be dealing with a civilization in transition from Type 1 to Type 2. The darkening of Tabby's Star is most likely a natural phenomenon, but as long as it is not clear where it comes from, the idea of a gigantic extraterrestrial solar power plant remains at least a fantastic (thought) possibility (Janjic 2017, pp. 57–58; Gerritzen 2016, pp. 157–164).

An additional idea that plays a role in the context of our considerations of various contact scenarios (see Chaps. 7 and 8) is this: humanity itself has been leaving a technosignature in space since 1895. Since then, it has been sending electromagnetic waves into space that technologically 'mark' the Earth (Janjic 2017, p. 48). The first stronger radio signals left Earth in the 1930s, propagated at the speed of light, and are therefore now over 80 light years away and have already reached several hundred exoplanets. Attentive extraterrestrials could therefore have already discovered us (Moore 2014, p. 250).

Now, based on these considerations, what can be said about the estimation of the factor f_c of the Drake equation? The answer is: *nothing at all.* Drake and his colleagues estimated that "[...] 10–20% of the intelligent civilizations would try to locate and communicate with alien civilizations" (Drake and Sobel 1992, p. 61). This value seems rather arbitrary—and it is. Nevertheless, the consideration that 10 percent of all intelligent extraterrestrial species with a technological civilization might have an interest in interplanetary communication and/or leave traces indicating their existence through the use of their technologies seems intuitively sufficiently plausible to us that we would like to cautiously endorse this estimate. This gives f_c a value of **0.1**.

3.7   The Lifespan of Extraterrestrial Civilizations

The reflections of Frank Drake and his colleagues on the question of how long the average duration of existence of extraterrestrial civilizations could be were recognizably shaped by the threatening scenario of a nuclear war between the superpowers USA and Soviet Union. Drake and Sobel write:

> We Earthlings already had the means to annihilate ourselves in one fell swoop. We achieved this ability almost at the same time as we became visible to the cosmos. Indeed, one of our Green Bank discussants, Phil Morrison, had worked on the Manhattan Project—had actually armed the second atomic bomb, dropped on Nagasaki on August 9, 1945, though he became an activist for arms control almost immediately. If the capability for total planetary destruction lay within the reach of every technologically advanced civilization, how likely were we to find anybody in the universe? (Drake and Sobel 1992, pp 61–62)

Apart from the fact that this consideration is highly anthropocentric and projects human aggression onto extraterrestrial civilizations, the human potential for self-extinction is now known to have multiplied, and in addition to the threat to humanity posed by nuclear weapons, other potential dangers have been added, such as biotechnology, nanotechnology, artificial intelligence, environmental degradation, and so on. Regardless, several other scenarios are possible that could bring an abrupt end to human civilization through no fault of its own. Consider, for example, massive volcanic eruptions, the impact of a large comet or asteroid, epidemics of aggressive pathogens, etc. In fact, there have already been several phases in the history of the Earth in which massive species extinctions occurred, triggered by climatic changes, meteorite impacts, volcanic eruptions, etc., or in which life on Earth as a whole was even threatened (Ward 2009). With regard to the probability of existence and lifespan of potential extraterrestrial civilizations, the so-called *Great Filter is* often spoken of in this context. The idea behind this is that on the way to a technological civilization that is capable of surviving over a longer period of time, a series of steps is required, one or more of which are so improbable that overall only very few or even only *one* civilization—ours—can be expected in the universe (Schulze-Makuch and Brains 2017, pp. 201–206). Assuming the Great Filter existed, the somewhat macabre question is: Have we already left it behind or is it still ahead? Moreover, the assumption of a Great Filter results in a paradoxical situation: the more evidence we find that extraterrestrial life is possible and even probable, while at the same time we have no evidence for the existence of an intelligent extraterrestrial civilization, the more likely it is that the Great Filter actually exists and is still ahead of us (Urban 2014).

The idea of the Great Filter is interesting, but ultimately a pure thought experiment that is highly presuppositional and provides little clue to the question of the potential survival time of an extraterrestrial civilization. The latter is also true when we try to extract clues to clarify this question from the evolutionary history of our own species: Science journalist Michael Shermer, based on the lifespan of 60 different advanced human civilizations, has calculated an average lifespan of about 420 years (Shermer 2002). This approach seems to make little sense, since it focuses on the lifespan of individual cultures within a species, but not on its *overall civilization.* For several decades (or even longer), humanity has already achieved a qualitatively and quantitatively unprecedented degree of interconnectedness between the various cultures or societies, so that it is justified to speak of a *world society* (Luhmann 1975), which is relatively insensitive to the 'withdrawal' of individual participating cultures (e.g. through political isolation as in North Korea or through war and destruction as in Syria). No or hardly any well-founded statements can be made about the concrete survival duration of human world society or of human civilization as a whole, although factors can be named that could drastically increase the probability of survival of human civilization. These include political-civilizational measures to curb violence and war, technological and scientific progress to protect environmental resources essential to survival, defence systems against the impact of large comets or asteroids, or the establishment of human colonies outside Earth. The most important factor in the long-term viability of human civilization, however, may be quite different: the transition to a *post-biological mode of existence.*

We will deal with the subject of post-biological civilizations in more detail at a later point (Sect. 10.1). We can only say this much here: human culture is already liberating itself step by step from its biological basis. The driving force for this seems to be a longing for *immortality* that is deeply seated in human beings. On the one hand, with the help of modern medicine and genetic engineering, it is attempting to extend the biological limits of the human body ever further, to make it healthier, more efficient and longer-lived. Some experts even assume that by 100 or 200 years from now the biological aging process can be completely stopped, and thus the limit of death will have been completely abolished (Harari 2017, pp. 35–46). The second path to immortality leads via technological development. The US futurologist Raymond Kurzweil assumes that computer-based artificial intelligence will soon have exceeded the capacity of the human brain and that in the course of this the moment of the *technological singularity* will occur. This refers to the point in time when artificial intelligences will be able to optimize themselves, which could lead to a rapid acceleration of further technological development and to an exponential or 'explosive' growth of artificial intelligence. According to Kurzweil, it is very likely that such super-intelligent systems would develop consciousness or human consciousness would be transferable to them— and they would be potentially immortal (Buchter and Straßmann 2013). In the course of their further self-optimization, such super-intelligent systems would most likely reduce their dependence on certain environmental factors to a minimum, could explore and colonize space as so-called *Von Neumann probes*[10] and could also continue to exist under conditions that exclude biological life.

Of course, such considerations are essentially science fiction so far, but they at least show potential development possibilities that could extend the expected lifespan of the human civilization or a civilization of artificial systems following it many times over. Should these future projections apply in whole or in part to any extraterrestrial civilizations, the consequence would be that they could potentially have a long lifespan or even have reached a state of immortality. The probability of such scenarios can admittedly not be estimated, but we do not want to exclude them by any means, therefore, to make a last bow to the Drake equation, we consider an average life span of extraterrestrial civilizations of 20,000 years to be a reasonable assumption and can thus determine the last factor: $\mathbf{L = 20.000}$.

---

[10] These are hypothetical, self-replicating space probes equipped with artificial intelligence, conceptualized by the mathematician John von Neumann (1903–1957).

3.8   Where Are They?

It should first be emphasized once again that the Drake equation is ultimately nothing but a numbers game, since, as shown, (especially with regards to the determination of the last factors) must for expediency be highly speculative. It is nevertheless of great use, since it not only makes possible interesting thought experiments, but also gives some structure to our search for extraterrestrial intelligences. After our determination of the individual factors, the calculation would look as follows:

$$N = 2.25 * 1 * 0.05 * 0.5 * 0.5 * 0.1 * 20,000 = \mathbf{56.3}$$

According to this calculation, we would have to deal with *about 56 technologically advanced extraterrestrial civilizations* in our Milky Way. It is clear that if the individual values of the factors are varied, the results also vary dramatically. For clarification, we have entered various scenarios with different optimistic and pessimistic estimates of the individual factors in the following Table 3.1 and inserted the previous calculation as a 'moderate' scenario in the middle:

Depending on which concrete values the individual factors of the equation assume, we would either not have to deal with a single further civilization in our Milky Way or, in an optimistic scenario, even with several hundred thousand of them. At the present time, we can only state: Both one and the other could be true. We simply do not know. It should be noted, however, that these calculations *only* apply to *our own galaxy*. This means that in order to arrive at a total value for the entire universe, the result must be multiplied again by a value that is astronomical in the truest sense of the word, namely the number of galaxies in the universe. Current estimates assume that there are at least as many galaxies in

Table 3.1 The number of extraterrestrial civilizations—different scenarios

Scenario	R	f_p	n_e	f_l	f_i	f_c	L	N
Conservative	1	0.1	0.01	0.1	0.1	0.01	1,000	0,0
Moderately conservative	1.5	0.3	0.05	0.1	0.25	0.1	5,000	0,3
Moderately	2.25	1	0.05	0.5	0.5	0.1	20,000	56
Moderately optimistic	3	1.5	0.5	0.7	0.7	0.2	50,000	11,025
Optimistic	3.75	2.5	1	1	1	0.5	100,000	468,750

Source own representation

Fig. 3.3 The Andromeda Galaxy. One of at least 100 billion galaxies in the universe. *Source* public domain figure[12]

the universe as there are stars in the Milky Way, i.e. at least 100 billion[11] (cf. for example Leibundgut 2011, p. 145). This in turn has the consequence that, even if the probability of the emergence of an intelligent extraterrestrial species within a galaxy may be quite low, there could nevertheless be millions of extraterrestrial civilizations in the entire universe (Fig. 3.3).

So if, for all we know today, it is quite likely that intelligent extraterrestrial civilizations exist somewhere in the universe, why haven't we heard about them yet? In short, where are they? Exactly this question was asked by the Italian physicist Enrico Fermi (1901–1954). The so-called *Fermi paradox* goes back to him. It is based on the assumption that every advanced technical civilization begins to colonize space at a certain point in time, which means that the Earth

---

[11] Based on the observation data of the Hubble telescope, some scientists even assume that there are more than a trillion galaxies (cf. Conselice 2016).

[12] Available online at: https://de.wikipedia.org/wiki/Andromedagalaxie#/media/File:And romeda_Galaxy_((Accessed: April 10, 2018).

must already have been colonized or at least explored by alien beings, if intelligent life had developed somewhere in the universe before that on Earth. But since we have not yet discovered any traces of an extraterrestrial civilization, this would indicate that it does not exist. Exemplary for this thought pattern are the considerations of the physicist and former ESA astronaut Ulrich Walter, in which he relates the Fermi paradox to our galaxy:

> If there were many ETIs in the ten billion year old Milky Way, some of them would have to have developed more advanced technologies in the past ten billion years than we have in the 4.5 billion years of our Earth history – whereas our own technology already allows us to colonize the Milky Way today. Thus, some ETIs should have already appeared on Earth – which is not the case. (There are at best dubious reports of UFOs, but these are hardly ETIs). And other explanations why ETIs have not appeared on our planet so far have turned out to be groundless. The conclusion, therefore, can only be: If no ETIs have appeared so far, then this can only mean that there are not many ETIs in our Milky Way. (Walter 2001, p. X)

The Fermi paradox makes clear how strongly also scientific statements about the chance of a contact to extraterrestrial civilizations are guided by all too human ideas about the motives of those extraterrestrials and especially also about their technical possibilities. Fermi's already mentioned basic idea that extraterrestrial civilizations would *inevitably* colonize space and therefore must have reached Earth long ago is obviously anthropocentric and therefore extraordinarily questionable. Colonizations in human history have usually been based on motives such as overpopulation, scarcity of resources, power-political interests, destruction of one's own environment, and so on. There is no compelling reason to assume that even one of these motives must guide the actions of an extraterrestrial civilization.

However, the Fermi paradox already 'limps' at a completely different point: the statement that we have not yet discovered any traces of extraterrestrial civilizations implies a more or less thorough previous *search for* the same. In other words:

> Italian physicist Enrico Fermi believed that if extraterrestrial life existed in our galaxy, we should have noticed it by now. But how could we have done this? Our attempts to listen in on certain areas of the galaxy on a few radio frequencies can't exactly be considered an extended search for life out there. (Moore 2014, p. 327)

Indeed, our active search for extraterrestrial intelligence is still in its infancy, even after decades of its existence. In any case, there can be no talk of a 'great eavesdropping attack' on the aliens. How our search for extraterrestrial intelligences has been conducted so far and which strategies, technical considerations and presuppositions are connected with it is the subject of the next chapter.

References

Anton, Andreas & Michael Schetsche. 2015. Anthropozentrische Transterrestrik. Zur Kritik naturwissenschaftlich orientierter SETI-Programme. *Zeitschrift für Anomalistik* 15: 21–46.

Benner,Stevan A., Alonso Ricardo & Matthew A. Carrigan. 2004. Is there a Common Chemical Model for Life in the Universe? *Current Opinion in Chemical Biology* 8: 672–689

Boutle, Ian A., Nathan J. Mayne, Benjamin Drummond, James Manners, Jayesh Goyal, F. Hugo Lambert, David M. Acreman & Paul D. Earnshaw. 2017. Exploring the Climate of Proxima B with the Met Office Unified Model. Astronomy & Astrophysics, March 1, 2017. https://arxiv.org/pdf/1702.08463.pdf.

Buchter, Heike & Burkhard Straßmann. 2013. Die Unsterblichen. Eine Begegnung mit dem Technikvisionär Ray Kurzweil und den Jüngern der ‚Singularity'-Bewegung. Zeit Online am 27.03.2013. http://www.zeit.de/2013/14/utopien-ray-kurzweil-singularity-bewegung.

Cassan, Arnaud, Daniel Kubas & Jean Philippe Beaulieu. 2012. One or More Bound Planets per Milky Way Star from Microlensing Observations. *Nature* 481: 167–169.

Cocconi, Giuseppe & Philip Morrison. 1959. Searching for Interstellar Communications. *Nature* 184: 844–846.

Davies, Paul. 1999. Vorwort. In *Nachbarn im All. Auf der Suche nach Leben im Kosmos*, Hrsg. Seth Shostak, 9–14. München: Herbig.

Davies, Paul. 2007. 'Are Aliens Among Us?' *Scientific American* 297 (6): 62–69.

Drake, Frank & Dava Sobel. 1992. *Is anyone out there? The Scientific Search for Extraterrestrial Intelligence.* New York: Delacorte Press.

Gerritzen, Daniel. 2016. *Erstkontakt. Warum wir uns auf Außerirdische vorbereiten müssen.* Stuttgart: Kosmos.

Harari, Yuval Noah. 2017. *Homo Deus. Eine Geschichte von Morgen.* München: C. H. Beck.

Heidmann, Jean. 1994. *Bioastronomie. Über irdisches Leben und außerirdische Intelligenz.* Berlin: Springer.

Herrmann, Dieter B. 1988. *Rätsel um Sirius. Astronomische Bilder und Deutungen.* Berlin: Der Morgen.

Hoerner, Sebastian von. 2003. *Sind wir allein? SETI und das Leben im All.* München: C. H. Beck.

Janjic, Aleksandar. 2017. *Lebensraum Universum. Einführung in die Exoökologie.* Berlin: Springer.

Joshi, Manoj. 2003. Climate Model Studies of Synchronously Rotating Planets. *Astrobiology* 3 (2): 415–427.

Kayser, Rainer. 2009. Spitzer und Hubble. Exoplanet mit organischen Molekülen. Astronews vom 21.10.2009.

Kennicutt, Robert C. & Neal J. Evans. 2012. Star Formation in the Milky Way and Nearby Galaxies. *Annual Review of Astronomy and Astrophysics* 50 (1): 531–608.

Koshland Daniel E. Jr. (2002). The Seven Pillars of Life. *Science* 295: 2215-2216.

Krauss, Lawrence. 2002. Zahlenspiele mit Außerirdischen. In *Auf der Suche nach dem Außerirdischen*, Hrsg. Tobias Daniel Wabbel, 26–36. München: beustverlag.

Leibundgut, Peter. 2011. *Ausserirdische und was Sie darüber wissen sollten*. Neckenmarkt (Österreich): Novum pro.

Luhmann, Niklas. 1975. Die Weltgesellschaft. In *Soziologische Aufklärung*, Bd. 2., ders., 51–71. Wiesbaden: VS Verlag für Sozialwissenschaften.

Martinez, Claudio L. Flores. 2014. SETI in the Light of Cosmic Convergent Evolution. *Acta Astronautica* 104: 341–349.

Mayor, Michel & Didier Queloz. 1995. A Jupiter-Mass Companion to a Solar-Type Star. *Nature* 378: 355–359.

Mayr, Ernst. 1995. Space Topics: Search for Extraterrestrial Intelligence. https://web.arc hive.org/web/20081115225902/http://www.planetary.org/explore/topics/search_for_life/ seti/mayr.html.

Meierhenrich, Uwe J., Guillermo M. Munoz Caro, Jan Hendrik Bredehöft, Elmar K. Jessberger & Wolfram H.-P. Thiemann. 2004. Identification of diamino acids in the Murchison meteorite. *Proceedings of the National Academy of Sciences of the United States of America* 101 (25): 9182–9186

Moore, Ben. 2014. *Da draußen. Leben auf unserem Planeten und anderswo*. Zürich: Kein & Aber.

Nagel, Thomas. 2014. *Mind and Cosmos. Why the Materialist Neo-Darwinian Conception of Nature is Almost Certainly False*. Oxford: University Press.

Petigura, Erik A., Andrew W. Howard and Geoffrey W. Marcy. 2013. Prevalence of Earth-Size Planets Orbiting Sun-Like Stars. *Proceedings of the National Academy of Sciences of the United States of America* 110: 19273–19278.

Pirschl, Julia and Michael Schetsche. 2013. Aus Fehlern lernen. Anthropozentrische Vorannahmen im SETI-Paradigma – Folgerungen für die UFO-Forschung. In *Diesseits der Denkverbote. Bausteine für eine reflexive UFO-Forschung*, Hrsg. Michael Schetsche und Andreas Anton, 29–48. Berlin: Lit-Verlag.

Robitaille, Thomas P. and Barbara A. Whitney. 2010. The Present-Day Star Formation Rate of the Milky Way Determined from Spitzer-Detected Young Stellar Objects. *The Astrophysical Journal Letters* 710 (1): L11–L15.

Scalo, John, Lisa Kaltenegger, Antígona Segura, Malcolm Fridlund, Ignasi Ribas, Yu. N. Kulikov, John L. Grenfell, Heike Rauer, Petra Odert, Martin Leitzinger, Franck Selsis, Maxim L. Khodachenko, Carlos Eiroa, Jim Kasting and Helmut Lammer. 2007. M Stars as Targets for Terrestrial Exoplanet Searches and Biosignature Detection. *Astrobiology* 7 (1), 85–166

Schetsche, Michael. 2005. Rücksturz zur Erde? Zur Legitimierung und Legitimität der bemannten Raumfahrt. In *Rückkehr ins All* (Ausstellungskatalog, Kunsthalle Hamburg), 24–27. Ostfildern: Hatje Cantz.

Scholz, Mathias. 2014. *Planetologie extrasolarer Planeten*. Heidelberg: Springer Spektrum.

Schulze-Makuch, Dirk and Luis N. Irwin. 2004. *Life in the Universe. Expectations and Constraints*. Heidelberg: Springer.

Schulze-Makuch, Dirk and William Bains. 2017. *The Cosmic Zoo. Complex Life on Many Worlds*. Cham: Springer Nature.

Selsis, Franck, James F. Kasting, Benjamin Levrard, Jimmy Paillet, Ignasi Ribas and Xavier Delfosse. 2008. Habitable Planets Around the Star Gl 581? *Astronomy & Astrophysics*. https://arxiv.org/pdf/0710.5294.pdf. Zugegriffen: 14. März 2018).

Sheridan, Mark A. 2009. *SETI's Scope: How the Search for ExtraTerrestrial Intelligence Became Disconnected from New Ideas about Extraterrestrials*. Ann Arbor, MI: ProQuest.

Shermer, Michael. 2002. Why ET Hasn't Called. Scientific American. https://michaelshermer.com/2002/08/why-et-hasnt-called/.

Shostak, Seth. 1999. *Nachbarn im All. Auf der Suche nach Leben im Kosmos*. München: Herbig.

Urban, Tim. 2014. The Fermi Paradox. http://waitbutwhy.com/2014/05/fermi-paradox.html.

Walter, Ulrich. 2001. *Außerirdische und Astronauten. Zivilisationen im All*. Heidelberg: Spektrum Akademischer Verlag.

Wandel, Amri. 2014. On the Abundance of Extraterrestrial Life After the Kepler Mission. *International Journal of Astrobiology* 14 (3): 511–516.

Ward, Peter. 2009. Gaias böse Schwester. *Spektrum der Wissenschaft* 11: 84–88.

Weyer, Johannes. 1997. Technikfolgenabschätzung in der Raumfahrt. In *Technikfolgenabschätzung als politische Aufgabe*, Hrsg. Raban Graf von Westphalen, 465–483. München: Oldenburg.

Zaun, Harald. 2006. Bewohnte Welten um Rote Zwergsterne? Telepolis (Online-Magazin). https://www.heise.de/tp/features/Bewohnte-Welten-um-Rote-Zwergsterne-3404750.html.

SETI—History, Methods and Presuppositions of the Scientific Search for Extraterrestrial Intelligences

4

4.1 SETI—An Idea and Its History

The scientific *search for extraterrestrial intelligence* (SETI), as we mostly perceive it today and also understand it culturally, began quite practically in 1960, when Frank Drake, as part of Project *Ozma* at the National Radio Astronomy Observatory in Green Bank (West Virginia, USA), began the systematic, albeit in several respects very selective, 'listening' of the universe for electromagnetic waves of artificial origin. Drake used the radio telescope of the observatory for his first tentative attempts to receive extraterrestrial radio signals, concentrated on the stars *Tau Ceti* and *Epsilon Eridani* and analysed, as suggested by Cocconi and Morrison (1959),[1] frequencies in the so-called *water hole*, i.e. at 1420 megahertz and a wavelength of 21 cm. The project was unsuccessful, but Drake and a few other scientists implemented with Project *Ozma* a search paradigm that is still valid today in the context of SETI programs and is pursued with great technical effort: the research-practical legitimacy of which is based on the basic assumption that other intelligent living beings actually exist in the universe that could be potential transmitters and receivers of radio signals (cf. for example Dick 1996, pp. 414–472; von Hoerner 2003, pp. 146–197; Sheridan 2009, pp. 11–31).

In 1961, Drake hosted the first SETI conference at the Green Bank Observatory, where he presented his famous equation and discussed it with colleagues. The conference participants consisted of scientists from a wide variety of disciplines, such as the astronomers Carl Sagen and Otto von Struve, the physicist Philip Morrison, the biochemist Melvin Calvin, and the neurophysiologist John Cunningham Lilly (von Hoerner 2003, pp. 151–152). In retrospect, the Green

---

[1] The paper, only two pages long, in the journal *Nature is* still considered the theoretical starting point of the scientific search for extraterrestrial radio signals (Zaun 2010, pp. 39–40).

© The Author(s), under exclusive license to Springer Fachmedien Wiesbaden Gmbh, part of Springer Nature 2023
A. Anton and M. Schetsche, *Meeting the Alien*,
https://doi.org/10.1007/978-3-658-41317-0_4

Bank Conference can be considered the initial spark for the scientific establishment of the SETI paradigm (Gerritzen 2016, p. 74). Drake and Sobel (1992, p. XIV) summarize the impact of the conference as follows:

> If Project Ozma failed to detect a sign of extraterrestrial intelligence, I did succeed in identifying our group at Green Bank as people who were committed to SETI. It also portrayed SETI to other scientists and the world at large for the first time as a legitimate and doable scientific endeavour. And the project stimulated activity among others who started our interest but had been afraid to act or had lacked the means to search.

In 1966, Carl Sagan and Josef Shklovsky published the highly acclaimed book *Intelligent Life in Universe*, which helped the cause of SETI to gain further popularity, but it was to take several more years before the project was institutionally on a secure footing (for the time being). In the early 1970s, NASA funded a kind of feasibility or implementation study for SETI (Project *Cyclops*). The study proposed using a large number of coordinated radio telescopes to search for radio signals from extraterrestrial intelligences at distances of up to 1,000 light years (Billingham 2014, pp. 4–6). For cost reasons, the specific proposals of the NASA study were not implemented, but they formed the basis for future SETI projects. In the years that followed, NASA funded the SETI program to the tune of several million dollars annually. Instead of building a new antenna network as envisioned in the Cyclops study, SETI used existing radio telescopes in NASA's *Deep Space Network to* search for extraterrestrial signals. In the various SETI projects of the next three decades, the number of frequency channels analyzed increased enormously. For example, the project *SERENDIP*[2] (Search for Extraterrestrial Radio Emissions from Nearby Developed Intelligent Populations) at the University of California at Berkeley (UC Berkeley) started in 1979 with a frequency analyzer with 100 channels, the follow-up project *SERENDIP II (from* 1985) already worked with more than 65,000 channels, SERENDIP *III* (from 1992) with 4 million and SERENDIP IV (from 1995) with more than 160 million channels (Werthimer et al. 1995, p. 293).

---

[2] The chosen abbreviation refers to the serendipity concept invented by Horace Walpole in the middle of the eighteenth century, which states that only those make extraordinary discoveries who are prepared to do so in terms of their thinking and working style (Holzhauer 2015, pp. 5–10). In the SETI projects, attempts were made for decades to make this principle usable—although it was often overlooked that *chance discoveries can* at best be mentally prepared, but not forced.

Although the SETI program had a comparatively small budget, initially about 2 cents per US taxpayer per year, its continued funding was repeatedly questioned by politicians. Finally, in 1993, government funding was cut off completely because the SETI project had failed to "bag a single little green fellow" (quoted after Shostak 1998, p. 160; see also Garber 2014, p. 33), as Senator Richard Bryan of Nevada, who was partly responsible for the cancellation of public funding, cynically remarked. This severe blow almost meant the end for the SETI project, but the SETI researchers managed to win several donors for the further search for extraterrestrial intelligences, so that the work could be continued with the help of private funds (Shostak 1998, pp. 161–162).

After the cancellation of public funding and the financial restructuring, the most ambitious project in the SETI program to date began: With Project *Phoenix,* a total of about 800 stars in a search radius of 200 light years were searched for extraterrestrial signals between 1995 and 2004. Project *Phoenix* examines.

> a total of two billion channels of each of its stellar targets, covering the microwave band between 1,000 and 3,000 MHz. It does so by listening in a block of 28 million simultaneous channels for five or ten minutes, then shifting the block up the radio dial and listening to another set of 28 million channels. It takes about a day's worth of listening and shifting to search one star over the two billion channels that the Phoenix astronomers think are the best for alien broadcasts. (Shostak 1998, p. 164)

In order to cope with the enormous amounts of data generated by listening to the universe for signals of extraterrestrial intelligence, an innovative and elegant solution was developed: In 1999, the *SETI@home* project was initiated, which enables anyone with an Internet-enabled computer to participate in the search for extraterrestrial intelligence. Using free software that activates automatically in screensaver mode, SETI data packets are downloaded from the University of Berkeley server to private computers and scanned for artificial signals. After computation, the data is automatically sent back to Berkeley. In this way a gigantic computer network was created, with the help of which huge amounts of data could be calculated for SETI (Wabbel 2002, pp. 78–79). The underlying principle of *distributed computing* has proved so successful that *SETI@home* is now a model for various projects in which large amounts of data have to be processed.

Thanks to an extremely generous donation of 100 million US dollars from the Russian entrepreneur Yuri Milner, the most extensive SETI project to date has been underway since 2015 with *Breakthrough Listen.* The goal of the 10-year research is to search more than one million stars and even alien galaxies for radio signals from extraterrestrial intelligences. In technical terms, *Breakthrough Listen* surpasses all previous SETI projects by far:

The 'Breakthrough Listen' project is expected to operate 50 times more sensitively than any previous SETI program, covering a 10 times larger section of the sky than before; it is expected to listen to a 5 times larger area of the radio spectrum simultaneously; and it is expected to do all this 100 times faster than ever before. (Fence 2015)

To cope with the enormous amounts of data generated by the project, it is again cooperating with the *SETI@home* initiative. Dan Werthimer, one of the co-founders of *SETi@home*, remarked euphorically in view of *Breakthrough Listen*: "This new SETI program puts everything we have done so far in the shade" (quoted from Zaun 2015).

In total, more than 120 different search programs have already been carried out since the beginnings of SETI research within the framework of Project *Ozma*. In none of the projects could a clearly artificial signal be detected so far. However, there have always been anomalies in the data, some of which could not be conclusively explained (Gerritzen 2016, pp. 85–87). A narrowband radio signal that was recorded in 1977 at the Big Ear radio telescope, as part of a SETI project, has gained some notoriety and has been controversially discussed ever since: the *Wow signal*.

4.2   The Wow Signal

When the US astrophysicist Jerry Ehman routinely went through the computer printouts of the signal inputs at the Big Ear radio telescope of August 15, 1977, he was so impressed by what was recorded from 11.16 p.m. onwards that he scribbled "Wow!" next to it. In doing so, he had unwillingly given this strange anomaly in the radio telescope's recording data a memorable name (Shostak 1998, pp. 166–167). The recorded signal from the direction of the constellation Sagittarius had several astonishing characteristics: It stood out very clearly from the cosmic background noise, its frequency was close to neutral hydrogen, and it was received for 72 s. The latter is significant in that 72 s is exactly the time it took the Big Ear radio telescope to move out of the signal's reception range due to Earth's rotation, making it very likely that it was a signal of interstellar origin—and not an aircraft, satellite, or the like (Zaun 2017) (Fig. 4.1).

At first glance, the Wow signal meets almost all the criteria for an extraterrestrial signal of artificial origin, but the problem is that it has so far remained a singular event: all attempts to record it again have failed. Moreover, the technology used to record the signal was not powerful enough to determine whether the

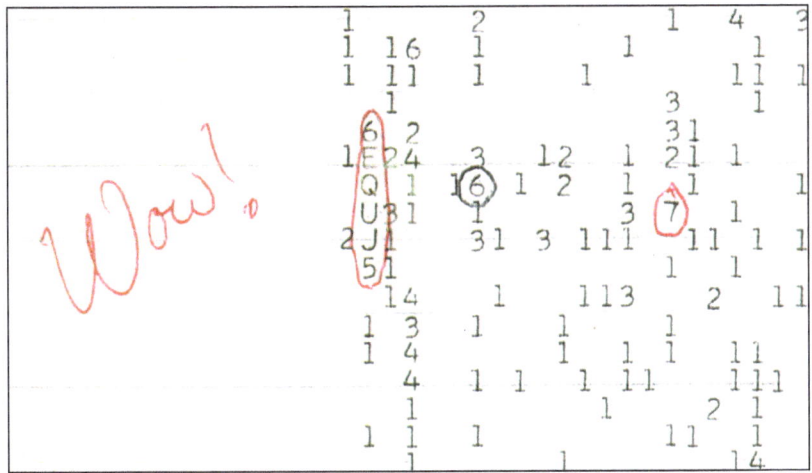

Fig. 4.1 The wow signal: evidence of an extraterrestrial civilization? *Source* public domain figure[3]

signal contained any kind of *content* in the form of a modulation. In the words of Jerry Ehman:

We collected one data point per channel every 12 seconds and collected a total of only 6 data points for Wow! Any variation of signal amplitude within the 12-second interval would not have been detected. The signal could have been varying in any of a variety of ways and we would not have seen it. Since the pattern of the 6 intensities followed our antenna pattern so well (with a correlation coefficient of between 99% and 100%, i.e., almost perfect), the signal falling on our telescope had an average value that did not change appreciably over the 72-second observing time. Saying that the average value didn't change does not tell you anything about the short- term variations in the signal. The signal could have been varying (modulated) at a frequency faster than once every 5 seconds (or 0.2 Hz, corresponding to half the data collection period) and we wouldn't have seen that modulation since our observatory was not equipped to detect such modulation. (Ehman 1998)

Ehman and his colleagues tested various possible explanations for the signal, such as planets, asteroids, comets, space probes, satellites, radio or television transmitters, etc., but were unable to identify its origin in this way (Gerritzen

---

[3] Available online at: https://de.wikipedia.org/wiki/Wow!-Signal#/media/File:Wow_signal.jpg.

2016, pp. 81–83). Attempts by other scientists to find an explanation for the wow signal have not yet yielded conclusive results either. In 2017, the US astronomer Antonio Paris published a paper in which he posits that the Wow signal could have been caused by the hydrogen cloud of a passing comet.[4] He cited comets *266P/Christensen* and *P/2008 Y2*, discovered in 2006, as possible candidates (see Paris 2017). Critics of this explanatory approach, on the other hand, object that these comets were not in the observation window of the Big Ear radio telescope at the time the Wow signal was recorded, and that the characteristics of hydrogen emissions from comets do not match the characteristics of the Wow signal in any way (Dixon 2017). Thus, the origin of the Wow signal remains unclear.

4.3   Anthropocentric Presuppositions

The presuppositions necessary for the SETI programs go far beyond the mere existence of intelligent beings beyond Earth: The civilizations projected in the context of SETI *must* possess a multitude of properties and abilities, without which most of today's SETI programs would simply be pointless. Some of the presuppositions of the SETI project have been criticized in recent years from the perspectives of developmental biology, communication science, or even philosophy. The tenor of this criticism is summarized by Sheridan (2009, p. 6):

> SETI searchers look through a very small keyhole: SETI searches for humanoid ETIs—which competent authorities think are unlikely—and does not look for the non-humanoid ETIs that are increasingly thought to be possible.

Indeed, the ideas held by many SETI researchers about basic characteristics of life and intelligence, as well as about the possibilities of interstellar communication, reveal an anthropocentrically colored picture of the extraterrestrials they seek. Thus, the dominant understanding of extraterrestrial intelligence in SETI programming is (inevitably) oriented almost exclusively toward humanoid capabilities and achievements, such as language, technology, ability and willingness

---

[4] Quite typical for the style of mass media reporting about the search for extraterrestrials is that immediately after the publication of Paris' essay a multitude of German print and online organs hurriedly reported that the wow-signal was now 'scientifically clarified'. Even a knowledge magazine like SPEKTRUM was quite lurid in its tone. An article entitled "Aus für Außerirdische" (Fischer 2017) is not able to distinguish the explanatory hypothesis of a single researcher from the prevailing expert opinion in the scientific community. The latter's criticism, some of which was quite sharp, followed Paris' essay on its heels (Schulze-Makuch 2017).

to communicate, and drive for research and expansion. Sebastian von Hoerner (2003, p. 103) formulated the conviction of those SETI scientists in an almost exemplary way:

> So for SETI, we hope elsewhere for advancing technology and science, especially for beings who are curious and talkative, who will even put a price on a thirst for knowledge and communication. And it seems that, as with us humans, this will require a strong individual, thinking intelligence.

Above all, the *technical-medial* communication capability of potential aliens is constitutive for the SETI program. For only if there are beings outside of Earth that have developed a technology that *matches ours* and are also willing to use it to communicate with other civilizations, does a search for extraterrestrial intelligence using radio waves make sense. Gerritzen (2016, pp. 62–66) summarizes what this means in concrete terms. An extraterrestrial civilization would have to:

- Shape metals into engineering tools and equipment and use these tools to process the metal into wires and cables;
- discover the electric charge, the electric voltage and electric fields;
- generate electricity and be able to call it up in large quantities;
- develop electrical grounding and substances with which electrical conductors can be insulated;
- build complex electronic components such as capacitors, transistors, diodes and processor chips from metals, ceramic and plastic materials and silicon;
- discover that a captured electromagnetic signal is amplified in, say, a coil of tightly wound copper wire;
- invent microphones and speakers to transmit speech or music;
- understand the physical properties of radio signals;
- develop physical-technical principles such as amplitude and frequency modulation;
- develop parabolic radio telescopes for the transmission of a radio signal and finally
- send out a strong radio signal at the right time on the right frequency specifically in the direction of our solar system.

We have traced these individual technical–technological steps in such detail in order to demonstrate how prerequisite-rich an establishment of contact with an extraterrestrial civilization is, according to the classical SETI paradigm. The assumptions about equivalent skills and interests of different extraterrestrial civilizations, which can be called '*contact optimism*', are supplemented in large parts of the SETI community by a '*communication optimism*', which assumes that civilizations willing to transmit are also able to encode their transmission in such a way that it can be understood by completely alien intelligences. The signals received would therefore first have to be recognised as artificial in the first place, and secondly they would then actually have to be decipherable. At the centre of the latter considerations today is usually the idea of mathematics as a kind of *universal language of* all intelligent beings. Already in the sixties of the last century, a 'cosmic language' called *Lincos* was developed, which would supposedly be understood by all civilizations, as long as they possessed certain basic mathematical skills (von Hoerner 2003, pp. 133–135; Ollongren 2010)—and, what is often forgotten to be mentioned, whose mathematics at least tend to correspond to ours.[5]

Another basic assumption of most SETI programs lies in the central importance assigned to certain physical properties of the element *hydrogen for* communication with extraterrestrials. From the beginning, SETI projects raised the question of which electromagnetic frequencies to focus on in the search for signals from extraterrestrial civilizations. The solution to the problem proposed by Cocconi and Morrison in 1959 was to determine the technical approach of SETI programs for decades: The search focused primarily on the 21 cm line of hydrogen as the base frequency (with possible halving and doubling). If one follows Sebastian von Hoerner's (2003, pp. 122–127) reasoning, hydrogen is not only the most abundant element in the universe, but must also be all too familiar to any intelligent living being, since it is, after all, one of the two chemical constituents of water, which is considered the universal building block of life. Every species interested in interstellar contacts would therefore, as it were by nature, choose that frequency for its passive search and for its active signals "which [should] be easily guessed by anyone already technically capable of contact and interested in it" (ibid., p. 123). This determination was only one of many consequences, which seemingly resulted from the once formulated premises of the SETI-paradigm as if by itself—with a more critical view probably derived from the world of imagination of terrestrial radio astronomers.

---

[5] That there is not even 'the mathematics' on Earth is shown by the research of the still quite young discipline of ethnomathematics (for an introduction see Ascher 1991). Against this background it seems to us rather doubtful to regard the today dominating terrestrial mathematics as in the original literal sense universal tool of all intelligent beings in the cosmos—and accordingly to assume it as quasi natural basis of understanding in the SETI projects.

Not only here exemplarily mentioned, but also other basic assumptions of the SETI paradigm show a considerable extent of anthropocentric bias when viewed from a historically and epistemologically sharpened perspective: The extraterrestrials 'projected' by most SETI programs to date originate from an *Earth-like* biochemistry based on water and carbon, their species was and is subject to *Earth-like* biological-evolutionary processes, it has undergone *similar* civilizational and technical developments as humanity, the individual beings are governed by very *similar* motives and therefore set out on a radio wave-based search for their 'brothers and sisters in space' at a certain point in their history.[6] In other words, the imagined alien is an almost flawless mirror image of those human scientists who had theoretically contoured the SETI programs and also practically implemented them. *The SETI researchers searched and search, in many cases until today, always only for themselves.* Who or what does not correspond to this (self-)image is not searched for and accordingly cannot be found. In a recent essay on the topic of the 'cosmic gorilla', the authors argue, alluding to the famous perceptual psychology experiment by US psychologists Simons and Chabris (1999), that the traces of an extraterrestrial civilization could be right 'under our noses', but that we have so far failed to register them because we have not paid attention to them or have been looking in the wrong place (De la Torre and Garcia 2018; Bohlmann and Bürger 2018, p. 166). This paper exemplifies that in recent years a rethinking has also begun among SETI researchers: More and more of the presuppositions that have been taken for granted for decades are now coming under scrutiny. From the perspective of the sociology of science, however, the question arises as to whether this is primarily due to inner-disciplinary insight or to the public pressure increasingly emanating from the continuing unsuccessfulness of all search projects. After almost 60 years of searching in the framework of more than 120 individual SETI projects, some of those involved are probably aware that a 'more of the same' approach is now unable to impress either the public or potential sponsors. Accordingly, alternative search strategies (such as optical SETI, the search for energy signatures and gigantic installations of alien civilizations, or even for extraterrestrial artifacts in our solar system[7]) have been gaining in popularity for years (Shostak 2006). The question remains, however, to what extent the new technical strategies are able to shake the foundations of the traditional SETI program.

---

[6] In many cases it was (and still is) taken for granted that the signals received originate from a biological species—and not from an artificial superintelligence. We will come back to this question in Chap. 10.

[7] The following Chap. 5 is dedicated to the search for extraterrestrial artifacts (SETA).

Critical objections against the presuppositions, models and technical strategies of the SETI paradigm are anything but new. As early as the mid-1960s, Soviet researchers, who had also intensively studied the possibilities of communication with extraterrestrial intelligences, identified major weaknesses in the traditional SETI paradigm (cf. for an overview Sheridan 2009, pp. 67–103). At a meeting in Byurakan, Armenia, the basic tenets of the SETI paradigm were critically discussed, particularly the remote contact and communication scenario prevalent among Western researchers. The central points of criticism of the Soviet scientists around Shklovskii, Kardashev and Ambartsumyan concerned on the one hand the probability of a successful identification of a signal of artificial origin, on the other hand the idea of a problem-free decipherability of the received extraterrestrial message. That the latter should be possible without an existing linguistic reference system was fundamentally questioned by Soviet linguists early on: If SETI-skeptics like Sukhotin (1971) or Panovkin (1976) are right and there is no possibility to translate isolated symbol systems, this means in its consequence that a communication via electromagnetic signals, as it is assumed in the western SETI-programs, is basically doomed to failure.

Besides this criticism of the contact optimism of SETI researchers, Soviet scientists also had fundamental doubts about the human-like nature of extraterrestrial intelligences assumed by the SETI paradigm. The connection between intelligence and the development of interstellar communication technologies, which was assumed to be necessary for the legitimacy of SETI programs at that time (and many of today's SETI programs) (according to this understanding—roughly speaking—intelligent is only the one who can communicate symbol- and technology-supported like humans; see Sheridan 2009, p. 109), was criticized in the following years not only by Soviet experts, but increasingly also in the West—for example from the ranks of evolutionary biologists and communication scientists (Sheridan 2009, pp. 103–128). One of the main arguments of the critics is that the development of human intelligence could well have been a unique process, i.e. the result of random evolutionary steps that are precisely *not* repeated elsewhere in the universe. Elsewhere and at other times, however, evolution may well have led to entirely non-humanoid forms of intelligence. As Sheridan (2009, pp. 141–166) points out, there now exists a plethora of quite plausible assumptions about the emergence of non-humanoid intelligences—a possibility, however, that has hardly been considered by the majority of (Western) SETI researchers to date. In addition, the highly presuppositional basic principles as well as the rather one-sided conceptualization of intelligence according to the model of human communication ability and technology isolate SETI research from many current debates about the possible constitution of extraterrestrial intelligence, which are conducted today in disciplines such as evolutionary biology, philosophy or sociology.

Furthermore, it should be noted: Fundamentally characteristic and problematic for SETI as a science is the fact that its object, extraterrestrial intelligence, is initially purely speculative (or hypothetical), and must be legitimized via auxiliary constructs, such as universally effective principles and mechanisms. Critical in this context is the fact alone that so far only one example, the Earth, can be used as evidence for the existence of planets on which intelligent living beings capable (and willing) of communication can be found. The core of the problem of extrapolation from one case, the *causa terra*, and thus the core of the SETI problem, is hit by the following question formulated by Engelbrecht (2008, p. 219): "Which of the factors that led to our existence are generalizable beyond Earth ('universal features') and which are specific to us ('parochial features')?".

An answer is pending—as long as evidence is found that can shed light on the existence and nature of (intelligent) extraterrestrial life. As long as this evidence is lacking, it seems to us clearly more reasonable (and in view of the considerable epistemological problems more appropriate) not to imagine extraterrestrial life exclusively as an extension of terrestrial conditions, as it is still frequently done in the context of SETI research, but also to include equally plausible non-humanoid (for instance also post-biological) extraterrestrials in the considerations and contact plans (Hövelmann 2009, pp. 179–181; Sheridan 2009, pp. 141–166).

4.4 General Communication and Understanding Problems

Until today, the main problem in thinking about communication with extraterrestrial entities is that we can hardly make justified statements about the—initially only imagined—counterpart. It seems quite possible that extraterrestrials in a—whatever kind of—contact with humans, prove to be alien to such a high degree that an understanding could succeed only with difficulty or even remain *permanently impossible*. Human understanding of aliens is based on.

> basic anthropological assumptions that make it possible to assume similar bodily needs, sensory possibilities, modes of perception, motivational states, coherent belief systems, etc. in the counterpart. (Schetsche 2004, p. 19)

We cannot unquestioningly assume these preconditions for communication with extraterrestrials. For example, an extraterrestrial civilization could be so far superior to us, or it could simply be so 'different' that we would not even be able to perceive the most vehement attempts at communication as such.[8] Although statements about very concrete communication possibilities must always remain speculative against this background, at least some basic considerations about the most abstract prerequisites of an information exchange between humans and extraterrestrials can be made on the basis of our current scientific knowledge.

Evolutionary biology[9] assumes today that the sensory channels of living beings develop according to the environmental conditions under which they exist. This means that, depending on the concrete conditions on their planet of origin, extraterrestrials will have adapted sensory channels and corresponding communication possibilities—and ones that by no means have to correspond to those of terrestrial land dwellers. For example, receptors (on Earth mostly in the form of so-called 'eyes') for electromagnetic radiation of certain wavelengths will only evolve evolutionarily if radiation of this frequency is present in the corresponding habitat in sufficient quantities to enable orientation in the environment (which is why many terrestrial animal species living in caves or in the deep sea do not possess them). The same applies to the senses of hearing and smell. On the other hand, creatures that have evolved on planets with different environmental conditions could have receptors that are—at least naturally—not available to us as humans: a sense for radioactivity, for magnetic fields or even for electricity. Already in some terrestrial creatures we find corresponding receptor organs, thus also sensory perceptions that remain cognitively *alien* to us humans despite all technical aids.

Since communication is directly tied to the sensory channels that make it possible, this means that the inhabitants of alien planets could use forms of communication that are unknown to us by virtue of our biological nature and that we are only able to simulate, if at all, with great technical effort (one thinks here, for example, of communication via tiny amounts of messenger molecules or radioactive substances). The problems described here intensify many times over if, in addition, one includes in the considerations post-biological secondary civilizations, whose development follows a logic unknown to us.

---

[8] This problem is referred to in the paper by De la Torre and Garcia (2018) mentioned earlier.

[9] The fundamental question of the applicability of the findings of terrestrial evolutionary biology to extraterrestrial civilizations is discussed in detail in Sect. 10.2 of this volume. (In accordance with the outcome of this later discussion, we take the liberty at this point of assuming the transferability, at least in principle, of basic assumptions of evolutionary theory).

With most of such (more or less) well imaginable *sense channels and communication modes*, the classical SETI research approach does not have to concern itself with more, as long as it favours—with doubtful reason[10]—a remote contact paradigm, according to which the communication with extraterrestrial civilizations will be possible exclusively by means of electromagnetic signals. If this basic assumption should be true, the only question would be whether an alien civilization from the extraordinarily broad spectrum would use exactly those electromagnetic frequencies which the terrestrial SETI researchers prefer (for the presupposition-rich reasons explained above) for communication attempts.

A collection of texts published by NASA in 2014 (Vakoch 2014) systematically deals with the prerequisites and limits of communication with extraterrestrial civilizations. The historian Ben Finney and the anthropologist Jerry Bentley, for example, compare the decoding of an extraterrestrial signal with the decoding of Egyptian hieroglyphics and the Mayan script. The latter caused great difficulties for centuries due to the destruction of almost all Mayan codices by the Spanish conquistadors, which is why Finney and Bentley are right to ask:

> The Maya case appears to undermine SETI scientists' hopes of actually translating the messages they are working to detect. If we have been unable to translate ancient human scripts without some knowledge of the spoken language they represent, what prospects do we have of being able to comprehend radio transmissions emanating from other worlds for which we have neither 'Rosetta Stones' nor any knowledge of the languages they encode? (Finney and Bentley 2014, p. 75).

Even if the aliens used a mathematical language, Finney and Bentley argue, it is by no means certain that we would be able to decipher the message. Thus, according to their argumentation, the historian Charles Étienne Brasseur de Bourbourg (1814–1874) was able to decipher the most important number symbols of the Maya early on and thus understand their calendar system, since it was based on mathematical principles—but the *texts* of the Maya could not be deciphered with it. Even if it were possible to find codes for physical, mathematical or formal logical connections within an extraterrestrial message, this would not mean that the message *as a whole would* be understood. The conclusion of Finney and Bentley is therefore (like that of the Soviet linguists decades earlier) skeptical:

---

[10] Again, to avoid too many duplications in the argumentation, we can only refer to a later chapter: In Sect. 7.2 we discuss, among other things, the preliminary assumption of the space-engineering unbridgeability of interstellar distances.

> We must think about the formidable prerequisites of deciphering extraterrestrial mes-
> sages and consider the possibility that whole domains of knowledge may remain
> opaque to us, despite our best efforts, for a very long time. If terrestrial analogues
> are to be employed in relation to SETI, then we should explore the wide range of
> human experience around the globe and not focus solely on familiar cases that appear
> to reinforce our most earnest hopes. (Finney and Bentley 2014, p. 77)

In the case of a non-medially mediated first contact, i.e. in the case of an
encounter with automatic exploration probes or even spaceships, on the other
hand, the question of the communication channels that can be used would be of
decisive importance. A physical direct contact, especially if it takes place on the
surface of a celestial body with atmosphere, allows the use of a wide spectrum
of communication forms: Sound waves, tactile vibrations, chemical messengers,
radioactive radiation, corporeal positioning in space, etc. However, some of these
signal forms could be so unusual from an earthly point of view that we humans
have not yet even devised appropriate technical sensors to transform these signals
into what we can see, hear or touch.

But even if we could determine the communication channel used and build
appropriate sensors, the serious problem of decoding the communicates remains.
According to all we know today, interpretative understanding of the other person
requires a minimum of common world experience and at least some compatible modes of world perception (Schetsche et al. 2009)—as shared by species
from similar habitats on the same planet (e.g. humans and wolves), but hardly by
beings originating from worlds with completely different environmental characteristics. The practical problems of communication that we are likely to encounter
when we meet an extraterrestrial species can hardly be imagined today, even with
all our imagination. Thus, from a human point of view, all that really remains
is the hope that 'the others' are also far ahead of us in terms of the theory and
practice of communication with beings from other worlds (Anton and Schetsche
2015, pp. 31–35).

4.5   A Brief Conclusion from a Sociological Perspective

The more than 120 SETI projects of the last decades are undoubtedly meritorious.
Like no other endeavor in space exploration, they have alerted a broad public to
the *possibility* that Earth may not be the only planet in our galaxy inhabited
by intelligent beings. The importance of this 'cosmological enlightenment' to

human self-perception[11] can hardly be overestimated. On the other hand, the SETI programmatic, as we had outlined above, is also responsible for a threefold narrowing of the professional and public debate on the subject of extraterrestrial intelligences:

1. the overwhelming part of the SETI experiments since the sixties of the last century reduces our idea of alien intelligences to *extraterrestrial radio astronomers.* The aliens whose radio messages we *can* receive resemble terrestrial SETI researchers like peas in a pod: They have developed a similar radio technology, follow identical lines of thought (for instance regarding that infamous 'waterhole wavelength'), and are also still in a similar technological development phase as Western societies on Earth (Bohlmann and Bürger 2018, p. 166). We think: It would have to be a freakish coincidence that all these (and some other) factors come together to make communication via radio waves over interstellar distances possible. Or our galaxy is hopelessly overpopulated, so that every century several new technological civilizations emerge at a suitable distance—some of which then also experiment with radio waves.

2. A second basic assumption, perpetuated for far too long, is that the signals received from foreigners can be decoded relatively easily and thus interpreted meaningfully. For decades the indications of especially Soviet linguists were ignored that there is basically no possibility to translate isolated symbol systems which are not based on a common practice of action (with direct communication and a reference system oriented to the material environment). It was not until recently that even Western SETI researchers admitted that their preliminary assumptions, regarding the decoding of alien signals were probably clearly too optimistic. The idea of a common 'intergalactic mathematics' and a 'universal' language based on it (which is only convincing at first sight) has manoeuvred research into a dead end for decades. Even a brief examination of the findings of the earthly social sciences with regard to the *problems of intercultural understanding of others* would have been able to put some convictions into perspective at an early stage (Bohlmann and Bürger 2018).

However, the highly improbable technical and linguistic presuppositions ultimately represent only the (in the strategic consequence: serious) consequences

---

[11] Here we are dealing with nothing less than the question of man's place in the cosmos, which has preoccupied mankind in the most diverse forms for thousands of years (see our remarks on this in Chap. 2).

of a *fundamental anthropocentric misunderstanding* to which the SETI program-matic is subject. Not least due to the (partly openly admitted) fascination for science fiction, especially US-American TV formats, alien images had taken root in the minds of many actors, which were rather dominated by the film-technical possibilities of the sixties and seventies than by scientific and (here almost more important: philosophical) reflections. The media image of *Vulcans* and *Klingons* was simply too seductive. Only when one manages to resist such striking media temptations does one's head become free for rational analysis: the real aliens are likely to be considerably stranger in every respect than we are able to imagine them today. This concerns not only detailed issues such as technology development and research strategies, but much more fundamentally their sensory channels, perceptual spaces, spatiotemporal orientations, and certainly the most basic ways of thinking.

As a consequence of this programmatic (and thus unfortunately also research pragmatic) misunderstanding, the search for extraterrestrials has for decades fol-lowed strategic paths that, when viewed critically, make it seem rather unlikely that they will lead to the desired success in the foreseeable future,[12] namely first contact with an extraterrestrial civilization. In the following chapter we will devote ourselves to an alternative search strategy, which in our view could complement SETI in a meaningful way: SETA—the search for extraterrestrial artifacts.

---

[12] In our opinion, one of the most serious misconceptions culminated in the dictum that we will never be able to encounter extraterrestrials directly because of the 'insurmountably large distances in the universe'. We will examine this fallacy more closely in Sect. 7.2. At this point, it is sufficient to note that this anthropocentric prejudice, despite all counter-arguments, however good they may be, still dominates both scientific and public debates today—which could lead to the social-psychological thesis that only the aliens, mentally moved to an almost insurmountable distance, seem to allow us to sleep peacefully at night (cf. Schetsche 2008, pp. 227–228). We do not even want to think through to the end the more radical connection thesis that some of those involved are unconsciously not at all concerned with establishing contact, but on the contrary with preventing it.

References

Anton, Andreas & Michael Schetsche. 2015. Anthropozentrische Transterrestrik. Zur Kritik naturwissenschaftlich orientierter SETI-Programme. *Zeitschrift für Anomalistik* 15: 21–46.

Ascher, Marcia. 1991. *Ethnomathematics – A Multicultural View of Mathematical Ideas.* Pacific Grove: Brooks/Cole Publishing.

Billingham, John. 2014. SETI: The NASA Years. In *Archeology, Anthropology and Interstellar Communication*, Ed. Douglas Vakoch, 1–21. Washington: National Aeronautics and Space Administration.

Bohlmann, Ulrike M. & Moritz J. F. Bürger. 2018. Anthropomorphism in the Search for Extra-Terrestrial Intelligence – The Limits of Cognition? *Acta Astronautica* 143: 163–168.

Cocconi, Giuseppe & Philip Morrison. 1959. Searching for Interstellar Communications. *Nature* 186: 670–671.

De la Torre, Gabriel & Manuel A. Garcia. 2018. The Cosmic Gorilla Effect or the Problem of Undetected Non Terrestrial Intelligent Signals. *Acta Astronautica* 146: 83–91

Dick, Steven J. 1996. *The Biological Universe: The Twentieth-Century Extraterrestrial Life Debate and the Limits of Science.* Cambridge: Cambridge University Press.

Dixon, Robert S. 2017. Statement Regarding the Claim that the „WOW!" Signal was Caused by Hydrogen Emission from an Unknown Comet or Comets. http://naapo.org/WOWCometRebuttal.html.

Drake, Frank & Dava Sobel. 1992. *Is anyone out there? The Scientific Search for Extraterrestrial Intelligence.* New York: Delacorte Press.

Ehman, Jerry R. 1998. The Big Ear Wow! Signal. What We Know and Don't Know About It After 20 Years. http://www.bigear.org/wow20th.htm#printout.

Engelbrecht, Martin. 2008. SETI. Die wissenschaftliche Suche nach außerirdischer Intelligenz im Spannungsfeld divergierender Wirklichkeitskonzepte. In *Von Menschen und Außerirdischen. Transterrestrische Begegnungen im Spiegel der Kulturwissenschaft*, Hrsg. Michael Schetsche und Martin Engelbrecht, 205–226. Bielefeld: transcript.

Finney, Ben and Jerry Bentley. 2014. A Tale of Two Analogues Learning at a Distance from the Ancient Greeks and Maya and the Problem of Deciphering Extraterrestrial Radio Transmissions. In *Archeology, Anthropology and Interstellar Communication*, Ed. Douglas Vakoch, 65–77. Washington: National Aeronautics and Space Administration.

Fischer, Lars. 2017. Aus für Außerirdische. SPEKTRUM online: News 06.06.2017. https://www.spektrum.de/news/aus-fuer-ausserirdische/1462193.

Garber, Stephen, J. 2014. A Political History of NASA's SETI Program. In *Archeology, Anthropology and Interstellar Communication*, Ed. Douglas Vakoch, 23–48. Washington: National Aeronautics and Space Administration.

Gerritzen, Daniel. 2016. *Erstkontakt. Warum wir uns auf Außerirdische vorbereiten müssen.* Stuttgart: Kosmos.

Hoerner, Sebastian von. 2003. *Sind wir allein? SETI und das Leben im All.* München: C. H. Beck.

Hövelmann, Gerd. 2009. Mutmaßungen über Außerirdische. *Zeitschrift für Anomalistik* 9: 168–199.

Holzhauer, Hedda. 2015. *Kriminalistische Serendipity – Ermittlungserfolge im Spannungsfeld zwischen Berufserfahrung, Gefühlsarbeit und Zufallsentdeckungen.* (Dissertation, Universität Hamburg, Fachbereich Sozialwissenschaften).

Ollongren, Alexander. 2010. On the Signature of LINCOS. *Acta Astronautica* 67: 1440–1442.

Panovkin, Boris Nikolaevich. 1976. The Objectivity of Knowledge and the Problem of the Exchange of Coherent Information with Extraterrestrial Civilizations. *Philosophical Problems of 20th Century Astronomy* (Moscow: Russian Academy of Sciences): 240–265.

Paris, Antonio. 2017. Hydrogen Line Observations of Cometary Spectra at 1420 MHZ. *Journal of the Washington Academy of Sciences 103 (2).* http://planetary-science.org/wp-content/uploads/2017/06/Paris_WAS_103_02.pdf.

Sagan, Carl, und Iossif Samuilowitsch Schklowski. 1966. *Intelligent Life in the Universe.* San Francisco: Holden-Day

Schetsche, Michael. 2004. Der maximal Fremde – eine Hinführung. In *Der maximal Fremde. Begegnungen mit dem Nichtmenschlichen und die Grenzen des Verstehens,* Hrsg. Michael Schetsche, 13–21. Würzburg: Ergon.

Schetsche, Michael. 2008. Auge in Auge mit dem maximal Fremden? Kontaktszenarien aus soziologischer Sicht. In *Von Menschen und Außerirdischen. Transterrestrische Begegnungen im Spiegel der Kulturwissenschaft,* Hrsg. Michael Schetsche und Martin Engelbrecht, 227–253. Bielefeld: transcript.

Schetsche, Michael, René Gründer, Gerhard Mayer & Ina Schmied-Knittel. 2009. Der maximal Fremde. Überlegungen zu einer transhumanen Handlungstheorie. *Berliner Journal für Soziologie* 19 (3): 469–491.

Schulze-Makuch, Dirk. 2017. Forty Years Later, SETI's Famous Wow! Signal May Have an Explanation. But the controversy continues (06.08.2017). https://www.airspacemag.com/daily-planet/forty-years-later-setis-famous-wow-signal-may-have-explanation-180 963628/.

Sheridan, Mark A. 2009. *SETI's Scope: How the Search for ExtraTerrestrial Intelligence Became Disconnected from New Ideas about Extraterrestrials.* Ann Arbor, MI: ProQuest.

Shostak, Seth. 1998. *Sharing the Universe. Perspectives on Extraterrestrial Life.* Berkeley: Berkeley Hills Books.

Shostak, Seth. 2006. The Future of SETI. Sky and Telescope online (19.06.2006). http://www.skyandtelescope.com/astronomy-news/the-future-of-seti/3/?c=y.

Simons, Daniel J. & Christopher F. Chabris. 1999. Gorillas in our Midst: Sustained Inattentional Blindness for Dynamic Events. *Perception* 28: 1059–1074.

Sukhotin, Boris Viktorovich. 1971. Methods of Message Decoding. In *Extraterrestrial Civilizations. Problems of Interstellar Communication,* Ed. S. A. Kaplan,133–212. Jerusalem: Keter Press.

Vakoch, Douglas A. 2014. Archeology, Anthropology and Interstellar Communication. Washington: National Aeronautics and Space Administration. https://www.nasa.gov/sites/default/files/files/Archaeology_Anthropology_and_Interstellar_Communication_TAGGED.pdf.

Wabbel, Tobias Daniel. 2002. Der Geist des Radios. In *S.E.T.I. Die Suche nach dem Außerirdischen,* Hrsg. Tobias Daniel Wabbel, 67–79. München: beustverlag.

Werthimer, Dan, David NG, Stuart Bowyer, und Charles Donnelly. 1995. The Berkeley SETI Program: SERENDIP III and IV Instrumentation. In *Progress in the Search for Extraterrestrial Life*, Ed. Seth Shostak, Astronomical Society of the Pacific Conference Series 74: 293–302.

Zaun, Harald. 2010. *SETI. Die wissenschaftliche Suche nach außerirdischen Zivilisationen.* Hannover: Heise.

Zaun, Harald. 2015. „Dieses neue SETI-Programm stellt alles Bisherige in den Schatten!" Telepolis am 21. Juli 2015. https://www.heise.de/tp/features/Dieses-neue-SETI-Programm-stellt-alles-Bisherige-in-den-Schatten-3374394.html?seite=all.

Zaun, Harald. 2017. Historisches SETI-Signal ohne Kosmogram. Telepolis am 15. August 2017. https://www.heise.de/tp/features/Historisches-SETI-Signal-ohne-Kosmogramm-3801610.html.

SETA—The Search for Extraterrestrial Artifacts

5

5.1 Alien Monoliths

In the second chapter we referred to some central works of science fiction literature, but deliberately refrained from mentioning one particular work that begins with a special scenario of an encounter with an extraterrestrial intelligence— Arthur C. Clark's *2001: A Space Odyssey*. Both the book and the film of the same name by US director Stanley Kubrick were released in 1968, and both are considered milestones of the science fiction genre. The book is the first part of a four-volume series dealing with humanity's contact with a highly advanced alien civilization (Clarke 2016). The film of the first in the series shines with innovative special effects, a brilliant screenplay, and an atmospheric depth and complexity that no other film in the genre had achieved before and rarely achieved after. Kubrick's film is, as Hurst (2004, p. 104) summarizes:

> a masterpiece of complex structural form and visual richness, it offers countless interpretive approaches while always remaining ambiguous, multilayered and inspiringly vague. As a result, 2001 has become one of the most discussed works in film history, and entire libraries can now be filled with interpretations and attempts at explanation.

The starting point of the story (in the film as well as in the book) is a group of prehumans or early humans three million years ago, who discover a *monolith* apparently deliberately placed by an extraterrestrial civilization. This produces a kind of leap in consciousness or intelligence in the early humans, as a result of which they are able to develop complex thoughts, make tools and weapons, and assert themselves against hostile groups of early humans. In short, the alien artifact provides the foundation for the development of human civilization. Three million years later, a *similar all-black monolith* is discovered by an exploration

A. Anton and M. Schetsche, *Meeting the Alien*, https://doi.org/10.1007/978-3-658-41317-0_5

mission on the Moon, buried in the floor of the crater *Tycho*, whose material properties clearly indicate that it is of artificial origin—and thus must be extraterrestrial. The exact nature, construction and possible functions of the monolith are completely unclear, but its presence proves in an almost provocative way that there was an extraterrestrial civilization that was far ahead of the human one in its development and that was or is still capable of depositing such an artifact on the Earth's satellite. As if this were not eerie enough, the monolith also emits a mysterious electromagnetic signal when it first comes into contact with sunlight.

The opening of *2001: A Space Odyssey* deliberately plays with human longings and fears in relation to potential extraterrestrial civilizations. This is enormously powerful, precisely because the staging of the encounter with an extraterrestrial civilization remains vague, unclear, and ambivalent. The black monolith stands symbolically for everything alien, inexplicable, inscrutable, which can mean both promise and threat, order and chaos. In this way, the book and the film succeed— quite in contrast to other science fiction narratives – "in presenting the alien as truly strange and almost incomprehensible" (Hurst 2004, p. 107).

The initial scenario of *2001: A Space Odyssey*—mankind discovers an extraterrestrial artifact—is rather neglected in the scientific discussion about potential extraterrestrial civilizations. The corresponding research perspective is called *SETA* (Search for Extraterrestrial Artifacts) in reference to SETI. The core idea of SETA is that technically advanced extraterrestrial civilizations have visited (or are visiting) our solar system or even the Earth itself (e.g., in the context of an exploration or research mission), possibly leaving behind traces of their presence that are identifiable to us. Such ideas may seem bold at first glance, but ultimately do not necessarily contain more or broader presuppositions than the classical SETI paradigm. Representatives of the SETA perspective like to refer to thought experiments according to which a high-tech civilization could succeed in exploring large parts of the Milky Way or even the entire Galaxy, within a relatively short time (at least by astrophysical standards). The basis for this would be automated probes controlled by an artificial intelligence, which could operate autonomously, reproduce themselves and thus explore space. Model calculations showed that such a mission of an extraterrestrial civilization could explore about one million stars within 10,000 years and the entire Milky Way in ten million years—and this at a speed of only one light year in 100 years (Gerritzen 2016, pp. 117–119). The proponents of the SETA perspective therefore claim that the search for extraterrestrial artifacts in our solar system should have an equal place alongside the classical SETI paradigm. We would like to take a closer look at the arguments associated with these considerations, and to this end we present in the following section the most important studies on the subject from recent decades.

5.2   SETA and the Fermi Paradox

One of the first to discuss the systematic scientific search for extraterrestrial artifacts was the mathematician, physicist and radio astronomer Ronald Newbold Bracewell. He considered it relatively pointless to search for radio signals from extraterrestrial civilizations (and in a very limited frequency range at that). His core argument was that even for.

advanced extraterrestrial civilizations [would be] simply too expensive to transmit radio signals into space over periods of years or decades with high energy expenditure, without knowing whether these would ever be received and 'heard' by another civilization. (Gerritzen 2016, p. 107)

It would be much more likely that alien civilizations would send out exploratory probes to explore space. Therefore, instead of looking for radio signals, it would make much more sense to look for extraterrestrial robotic probes. These considerations lead directly back to the *Fermi paradox* (we had introduced it in Chap. 3): if there are such alien exploration probes, where are they?

One proposed solution to the Fermi paradox comes from the year 1973 by John A. Ball and is called the *zoo hypothesis*. According to Ball's conviction, it is statistically very likely that civilizations exist here and there in our galaxy that are technically far ahead of ours. In his view, this also makes it likely that one or more of these technically advanced civilizations have already discovered our solar system, the Earth, and also *us. The* fact that we humans have not yet noticed anything about it (Ball explains), is because the extraterrestrial space-faring civilizations keep mankind *deliberately* in ignorance about their existence—be it in order to not disturb the development of the terrestrial civilization, or in order to be able to observe and examine it better. In Ball's words: "The zoo hypothesis predicts that we shall never find them because they do not want to be found and they have the technological ability to insure this" (Ball 1973, p. 349).

A few years later (1977), astronomers Thomas B. H. Kuiper and Mark Morris first discuss the classic SETI strategies for identifying extraterrestrial radio signals in an article in *Science*. They emphasize that in the search for extraterrestrials it is crucial to make only the *most necessary* preliminary assumptions regarding the capabilities and behaviors of the aliens. They therefore explicitly criticize the assumption that crossing interstellar distances by space probes or even 'manned' spacecraft is entirely impossible:

The contention of this article is not that the practice of interstellar travel is an inevitability for all technologically advanced civilizations, but that the probability is high enough that, given a modest number of advanced civilizations, at least one of them will engage in interstellar travel and thus colonize the galaxy. (Kuiper and Morris 1977, p. 616)

As a 'travel-technical' explanation they put forward, among other things, the possibilities that aliens have a significantly higher life expectancy than humans, that they use generational spaceships or that the crew could spend a large part of the journey in a kind of cold sleep. Accordingly, they believe it is possible that our solar system has already been visited by aliens, or at least studied by unmanned space probes. In their opinion, there are good reasons for the assumption that alien civilizations try to keep their presence in our solar system secret. In a short appendix to the article (Appendix A), they discuss in this context the risk of 'culture shock' when a culture still bound to its planet is confronted with a superior space-active civilization. In their view, this could be one reason why extraterrestrial civilizations have not yet made contact with us, or have not made themselves known to us, even though they are present in our solar system. The authors add:

In many cases, however, the needs of the aliens could be satisfied without undue impact on our civilization. The removal of rare elements or chemicals, of genetic material, or of samples for biological or psychological studies (including even an occasional human) could be effected with no more attention from us than a UFO article or a missing person's report. To establish that avoidance of open contact is not the most likely alien behavior, one would need to identify a resource that does not fall into this category. [...] There remains the possibility that members of an extraterrestrial society might choose limited contact without offering their store of knowledge. They might wish to do this (i) as part of an experiment to gauge the reaction of our society, (ii) in an attempt to stabilize terrestrial civilization to prevent an impending crisis of self-annihilation, or (iii) to plant selected information in order to stimulate our evolution in some preferred direction. In none of these cases can it be concluded that contact would necessarily occur in an overt way, so that we would immediately recognize it as such. (Kuiper and Morris 1977, p. 620)

However, in their view, these and similar questions go beyond the primary jurisdiction of physics and astronomy, so the authors suggest that other scientific disciplines should contribute their expertise to the debates about SETI and extraterrestrial civilizations.

In a seminal 1983 article for the *Journal of the British Interplanetary Society,* physicist Robert A. Freitas explains why he believes it makes sense to search for extraterrestrial artifacts in our solar system. Thereby he formulates the *artifact hypothesis*, which is central for the further discussion in this field, according to which an advanced extraterrestrial civilization would with high probability have developed a program for the exploration of interstellar space and would have sent out research probes or similar for this purpose: "A technologically advanced extraterrestrial civilization has undertaken a long-term program of interstellar exploration via transmission of material artifacts" (Freitas 1983, p. 501). According to Freitas, such an extraterrestrial research or exploration program could explore the entire galaxy in a relatively short period of time with the help of self-reproducing probes ("Von Neumann probes"):

> Self-replicating or self-growing probe factories need only produce a dozen or fewer offspring in each target star system to explore the entire Galaxy in less than a dozen generations, requiring $10^2 - 10^3$ years for completion of one generation at each site. This is $10^{-6} - 10^{-7}$ the age of the Earth, an improbably small observational window. (Freitas 1983, p. 502).

If this thesis should be true, he continues, there would be a chance to find corresponding material artifacts in the solar system, if one only searched for them with the corresponding technical effort. Based on the necessarily advanced technical capabilities of a civilization capable of sending space probes to other solar systems, he puts forward the following follow-up thesis:

> Artifacts not intended to be found will not be found. For instance, in one scenario the probe imperfectly camouflages itself with the motive of providing a thresholding test of the technology or intelligence of the recipient species, which test must be passed before communication with the device is permitted. [...] Thus only those classes of artifacts not subject to a policy of perfect concealment can be observed by us. (Freitas 1983, p. 501)

The efforts of extraterrestrials to avoid their detection by the observed seems to him a rather elegant solution to the Fermi paradox. Assuming that the Earth represents the most complex and interesting environment in our solar system, Freitas also assumes that extraterrestrial probes in our solar system would concentrate on observing the Earth and would be found at correspondingly suitable observation sites (i.e. in a more or less wide orbit around the Earth or on the Moon).

Basically, Freitas distinguishes four variants of extraterrestrial artifacts that could possibly be found in our solar system:

(1) *Astro-engineering*: This is the possibility that one or more extraterrestrial civilization(s) have technically manipulated celestial bodies in our solar system in some way.

(2) *Self-replicating artifacts*: Freitas is thinking here of autonomously operating robotic probes that reproduce themselves, for example, by setting up manufacturing factories, mining raw materials, etc.

(3) *Passive artifacts*: Passive artifacts can be, for example, leftovers ('garbage'[1]) from exploration missions, but also abandoned buildings, monuments, left behind measuring stations or deliberately left messages.

(4) *Active artifacts*: This refers to active probes or other observational technology that monitor the solar system and may send information back to their 'builders'—or even interact with species they discover (Freitas 1983, pp. 501–503).

Some years later (1987), the physicist and meteorologist James W. Deardorff discusses in a very fundamental way the most different solutions for the Fermi paradox. His proposals range from the possibility that we are simply alone in the universe to the *embargo hypothesis*, according to which technically advanced extraterrestrial civilizations would deliberately avoid any contact with humanity. The basic idea is that there could be some kind of agreement among highly developed extraterrestrial civilizations regarding a 'non-interference ethic' that prohibits them from influencing technically and ethically less developed civilizations and thus pretends to keep them, as it were, in a 'nature preserve' from alien influence. Deardorrf even considers this explanation more plausible than contact in the sense of the SETI paradigm. He writes at the beginning of his essay:

> The embargo or quarantine hypothesis for explaining the 'Great Silence' is reviewed and found to be more plausible than the view that, at most, we might expect radio messages from some distant star. The latter hypothesis is shown to be compatible with extraterrestrial technologies only a few hundred years in advance of your own, whereas the embargo hypothesis more reasonably infers that they should be tens of thousands of years in advance and in control of any contact with humanity. (Deardorff 1987, p. 373)

---

[1] The idea that aliens, after leaving the solar system, simply leave their rubbish here, can be found in the SF novel *Picnic by the Wayside by* the brothers Arkadi and Boris Strugazki (German: 1975, Russian original: 1971). The novel became known to a wider public through the film adaptation *Stalker* (by Andrei Arsenyevich Tarkovsky, USSR 1979).

Deardorff is particularly interested in the possibility that the Earth is now being systematically monitored by extraterrestrials—using such advanced technology that it is currently *impossible for* us to detect the observers and their instruments. In this context, he even addresses the thesis that at least some of the UFO sightings of recent decades *could* be interpreted in such a context:

> However, Sturrock noted that if covert or indefinite evidence of such visitations existed, that fact would be important to take into account. He therefore regarded the tens of thousands of screened reports of UFOs occurring in the past 40 years as possibly relevant in any discussion of ETIS. If so, then for compatibility with the embargo hypothesis individual UFO appearances would have to be governed by the same rules as the hypothesised embargo against Earth: no UFO evidence so definite to scientists as could cause the embargo to 'break' would be permitted. For further compatibility, description of UFO behaviour should disclose signs of intelligence, highly advanced technology, and generally ethical actions. (Deardorff 1987, pp. 377–378)

In an essay in the *Journal of The British Interplanetary Society* (1998), the Ukrainian radio astronomer and astrophysicist Alexey V. Arkhipov calls for a systematic search for extraterrestrial artifacts in our solar system—especially on the moon, since here, according to Arkhipov, would be very favorable conditions for the placement of an extraterrestrial observation probe. In his opinion, it could also make sense to search for 'waste' from extraterrestrial space missions in the solar system and on Earth. Arkhipov's approach thus bears a certain resemblance to the paleo-SETI hypothesis—with the crucial difference, however, that the author does not assume a priori that there are such legacies on Earth, but rather advises an open and unprejudiced search for possible evidence of extraterrestrial visitation(s). His conclusion:

> There are interesting nonclassical SETI possibilities which look more effective and promising than the conventional search for radio/laser signals. Unfortunately, new approaches conflict with the mental habits of astronomers, geologists and geochemists in studying natural formations and processes. The habit factor leads most specialists to an a priori rejection of search for alien artifacts on the surface of the Moon and the Earth. (Arkhipov 1998, p. 184)

In a 2011 paper for the *Journal of the British Interplanetary Society*, planetary scientist Jacob Haqq-Misra and astrophysicist Ravi Kumar Kopparapu discuss the possibility of identifying extraterrestrial artifacts in our solar system using observation technology available today or in the near future. Their initial hypothesis is that in principle there is a good chance of finding such artifacts in our solar system—at least if extraterrestrial civilisations pursue similar research strategies as mankind today. As part of their consideration of various proposed solutions to the Fermi paradox, the authors also address the zoo hypothesis (Ball 1973, see above), according to which the existence of Earth and its inhabitants is known to extraterrestrial civilizations, but there is a prohibition—of whatever kind and enforced—against revealing themselves to humanity. The two researchers are particularly interested here in the special case of Earth and the solar system being systematically monitored by automatic probes *within the framework of* such a ban on contact. In their view, this is quite conceivable because a highly advanced extraterrestrial technology could make it impossible for mankind to detect corresponding surveillance probes at the current state of its technical development. But even without passive or active safeguards against detection, the small size of extraterrestrial probes alone (the authors assume they would be only between one and a few meters long for efficiency reasons) could make them extraordinarily difficult to detect—at least as long as they are not systematically searched for. "Thus, extraterrestrial artifacts may exist in the Solar System without our knowledge simply because we have not yet searched sufficiently" (Haqq-Misra and Kopparapu 2011, p. 4). Consequently, Haqq-Misra and Kopparapu, analogous to Arkhipov, propose the development of a research program that systematically searches for extraterrestrial artifacts within our solar system.

Finally, in a paper for *Acta Astronautica* (2012), cosmologist and astrobiologist Paul Davies and his student Robert Wagner strongly call for looking for extraterrestrial artifacts as part of mapping the surface of our Earth's moon. The background of the proposal is the unsuccessfulness of previous SETI strategies and the increasingly obvious need to develop alternatives. In general, the two authors propose to use data from astronomy, biology, geology and planetary sciences that have already been collected or will be collected in the near future to supplement the existing search strategies in order to obtain clues about the existence of extraterrestrial civilizations by other means. Their specific idea is to use photographic data from the ongoing Lunar *Reconnaissance Orbiter mission* (launched in June 2009) to search the lunar surface for man-made structures of any kind—be they remnants of automated probes or even interstellar expeditions. They write:

> Such an artifact might originate in several ways, for example, discarded material from an alien expedition or mining operation, instrumentation deliberately installed to monitor Earth, or a dormant probe awaiting contact (a variant on the message-in-a-bottle theme). Alien technology might also manifest itself in mining or quarrying activity, or even construction work, traces of which might persist even after millions of years. (Davies and Wagner 2012, p. 2)

The Earth's moon seems to them to be particularly well suited for such a search because, due to the slowness of erosion processes and the extensive absence of tectonic processes, corresponding artifacts (or more generally: artificial structures) are not only preserved there over extremely long periods of time, but their signatures (such as radioactivity or magnetic fields) also stand out particularly clearly from the surroundings.[2] Because of the immense periods of time that have passed since the formation of stars and planetary systems, they consider it possible that corresponding visits have already taken place in the distant past (perhaps even hundreds of millions of years ago). In their paper, they discuss not only the detectability of various types of artificial structures or objects, but also the question of what terrestrial scientists today might do with such remnants. The details on this are unnecessary at this point, as are the details on how the data of the Lunar Reconnaissance Orbiter (or other space probes) can be scanned for corresponding 'anomalies' with minimal effort.

5.3 Conclusions

Let us first summarize the alternative search concept SETA again from an exosociological perspective:

(1) SETA projects do not represent a fundamental alternative, but a *supplement to* classical SETI research.
(2) They do not share the limiting presupposition that bridging interstellar distances (whether by space probes or even spacecraft) would be entirely impossible.
(3) Included in the proposed search are legacies of extraterrestrial exploration missions as well as still active extraterrestrial space probes in our solar system.
(4) The search strategies are of necessity oriented to anticipated research and observation missions, as they would be projected by a future terrestrial exploration of distant solar systems.

---

[2] Similar to Foster (1972) long before them, the authors distinguish four types of evidence for extraterrestrial activity that might be found on the Moon: (1) artifacts with an explicit message for Earth's inhabitants (or other spacefaring species), (2) observational instruments or their remnants, (3) debris left behind by interstellar expeditions, and (4) changes to geologic structures due to the study or mining of certain lunar materials.

(5) The more advanced the potential extraterrestrial technology is, the lower the chance is estimated to be to be able to find alien spacecraft etc. in our solar system—this applies all the more if they are deliberately protected against detection.

(6) Despite these limitations, the search for corresponding extraterrestrial artifacts in our solar system, and even on Earth itself, appears to be quite promising.

Although SETI and SETA need not be mutually exclusive, strategically the latter represents one of several alternatives to classical radio astronomical listening strategies.[3] The central conceptual difference between the two programmatic approaches is that SETA does not recognize the anthropocentric axiom that guides research in all SETI endeavors. It claims, (we have mentioned it several times) that even civilizations technologically only thousands of years ahead of Earth's are incapable of bridging interstellar distances with their spacecraft. Abandoning this presupposition, as the SETA program does, opens up a whole new set of possibilities for proving that humanity is not alone in the universe. In particular, by decoupling the success of the search from the necessity of the temporally parallel (or, depending on the distance, temporally staggered) existence of various civilizations,[4] the probability of proving the existence of alien intelligence increases significantly. It is now possible to discover evidence of alien civilizations whose technological heyday lies hundreds of thousands of years in the past and which have perhaps long since ceased to exist.

The omission of that presupposition, however, generates (we mentioned it above) a possibility space in which there is room for speculation about past or even present visits of extraterrestrial intelligences to Earth—these are, as we will show in Chap. 11 of this volume, scientifically rather undesirable questions today. In our view, this is one of the reasons why SETA projects are viewed with continued skepticism by many radio astronomical SETI researchers. Another reason may be the *disciplinary competition for* public attention and for financial resources that are not independent of it. It is striking that almost all ideas for SETI projects come from the field of radio astronomy, whereas the basic SETA idea

---

[3] Another concept, for example, asks whether giant artificial structures built by highly advanced civilizations can be discovered orbiting alien stars (see Zackrisson et al. 2018 for an example).

[4] As we had already explained in the previous chapter, this is a major problem of all SETI projects: In a galaxy that is many billions of years old, different civilizations must be in the same technological development phase at exactly the right time in order to be able to communicate with each other.

comes from experts in very different research fields who do *not* themselves run
SETI projects. To reconstruct the exact connections here would be rather the task
of science research[5] than of exosociology—it is clear, however, that with regard
to the research resources raised by the state or privately, the traditional SETI
projects are still very much ahead. We are not in a position to predict whether
this will change in the foreseeable future. In view of the fact that only very few
publications on the subject have appeared recently, we have the impression that
the discussion is currently in danger of dying out. We attribute the fact that the
SETA research perspective has not been able to establish itself scientifically since
its inception to three causes:

(1) The presence of an extraterrestrial artifact in our solar system (or even on
    Earth) would imply that the corresponding extraterrestrial civilization is capa-
    ble of bridging interstellar distances, which the majority of experts—namely
    many astrophysicists—consider highly unlikely, occasionally even impossi-
    ble. From a sociological point of view, SETA still represents a heterodox
    search strategy that struggles for recognition.
(2) Some scientists mentally place SETA projects in the vicinity of UFO research
    and the paleo-SETI hypothesis, both of which are equally accused of unseri-
    ousness and are therefore systematically avoided—also out of fear of loss of
    academic reputation.[6]
(3) Finally, the mass media dominance of the remote-contact paradigm prop-
    agated by SETI has also contributed to the fact that the question of
    extraterrestrial artifacts has long been relegated to the background. Public
    attention is so significant in this research context because most projects are
    funded by private donors. This is probably why delegitimization strategies
    are occasionally found in SETI research to prevent scarce funds from flowing
    into alternative search programs.

However, all these are no reasons that could legitimize a blanket rejection of
a search for extraterrestrial legacies on Earth or in our solar system. On the
contrary: on closer inspection, the search for extraterrestrial artifacts, at least in
our estimation, proves to be no less plausible and also no more presuppositional
than the search for extraterrestrial radio signals.

---

[5] A critical reconstruction of the history of the early SETI projects was provided by the
sociologist of science Daniel Ray Romesberg (1992)—corresponding research today would
probably fall under the jurisdiction of the so-called "Science and Technology Studies".
[6] See our detailed discussion in Chap. 11 of this volume.

Conclusion: Largely unnoticed by the public and only rarely scientifically received, a few scientists in recent decades—sometimes more, sometimes less systematically—have dealt with the possibility of obtaining certainty about the existence of extraterrestrial civilizations other than through classical SETI strategies. The central idea is that somewhere in our solar system, perhaps not so far from Earth, material remains of an extraterrestrial civilization could be found—be it remnants of long-past exploration missions or still active research and observation probes of extraterrestrial origin. In the context of their considerations, they have at the same time proposed a whole series of innovative solutions to the Fermi paradox. Whether one agrees with these ideas or not, it remains to be noted that there is—outside the radio-astronomically influenced leading paradigm—a whole series of other very serious proposals for a search for extraterrestrial intelligences. And the longer the various SETI projects remain unsuccessful, the more such alternative search strategies will become the focus of scientific, but also public interest.

References

Arkhipov, Alexey V. 1998. Earth-Moon System as Collector of Alien Artefacts. *Journal of the British Interplanetary Society* 51: 181–184.

Ball, John A. 1973: The Zoo Hypothesis. *Icarus* 19: 347–349.

Clarke, Arthur C. 2016. 2001: *Odyssee im Weltraum – Die komplette Saga*. München: Heyne.

Davies, Paul & Robert Wagner. 2012. Searching for Alien Artifacts on the Moon. *Acta Astronautica* 89: 261–265.

Deardorff, James W. 1987. Examination of the Embargo Hypothesis as an Explanation for the Great Silence. *Journal of the British Interplanetary Society* 40: 373–379.

Foster, G. V. 1972. Non-Human Artifacts in the Solar System. *Spaceflight* 14: 447–453.

Freitas Robert A. Jr. 1983. The Search for Extraterrestrial Artefacts (SETA). *Journal of the British Interplanetary Society* 36: 501–506.

Gerritzen, Daniel. 2016. *Erstkontakt. Warum wir uns auf Außerirdische vorbereiten müssen.* Stuttgart: Kosmos.

Haqq-Misra, Jacob & Ravi Kumar Kopparapu. 2011. On the Likelihood of Non-Terrestrial Artifacts in the Solar System. *arXiv*: 1111.1212v1.

Hurst, Matthias. 2004. Stimmen aus dem All – Rufe aus der Seele. Kommunikation mit Außerirdischen in narrativen Spielfilmen. In *Der maximal Fremde. Begegnungen mit dem Nichtmenschlichen und die Grenzen des Verstehens*, Hrsg. Michael Schetsche, 95–112. Würzburg: Ergon.

Kuiper, Thomas B. H. & Mark Morris. 1977. Searching for Extraterrestrial Civilizations. *Science* 196: 616–621.

Romesberg, Daniel Ray. 1992. *The Scientific Search for Extraterrestrial Intelligence: A Sociological Analysis.* Ann Arbor: UMI Dissertation Services.

Zackrisson, Erik, Andreas J. Korn, Ansgar Wehrhahn & Johannes Reiter. 2018. SETI with Gaia: The Observational Signatures of Nearly Complete Dyson Spheres. https://arxiv.org/abs/1804.08351.

Calls in the Dark Forest—Risky Communication Attempts

6

6.1 Messages to Extraterrestrials

On March 3, 1972, an Atlas rocket carrying the *Pioneer 10* space probe was launched at the *Cape Canaveral* rocket launch site in the U.S. The goal of the mission was to explore the asteroid belt and Jupiter, and then traverse outer solar system space. In November 1973, Pioneer 10 reached the Jupiter system, in 1976 the probe passed the orbit of Saturn, and in 1979 that of Uranus. In 1983, Pioneer 10 finally became the first man-made object to leave the space of our solar system and has since been travelling in interstellar space towards the star *Aldebaran* in the constellation of Taurus, which the probe will reach after about 2 million years of travel. The Pioneer 10 probe remained the furthest human object from Earth until February 1998—when it was 'overtaken' by the *Voyager 1* probe. Pioneer 10 provided numerous scientific discoveries and the first detailed images of Jupiter and its moons. The mission was a complete success from a technical and scientific point of view and was also distinguished by a special feature: Attached to the spacecraft is a gold-coated aluminium plate containing a message to potential extraterrestrial intelligences. Although the likelihood of the Pioneer spacecraft one day being discovered by an extraterrestrial civilization was considered extremely low, NASA agreed to attach the metal plaque to the spacecraft—primarily because of the presumed positive public impact—and commissioned Frank Drake and Carl Sagan to create a message from humanity to potential extraterrestrial discoverers of the Pioneer spacecraft. The message contains various symbolic images, including a schematic representation of a man and a woman and of Earth's position in our solar system (Drake and Sobel 1994, pp. 253–261) (Fig. 6.1).

© The Author(s), under exclusive license to Springer Fachmedien Wiesbaden GmbH, part of Springer Nature 2023
A. Anton and M. Schetsche, *Meeting the Alien*,
https://doi.org/10.1007/978-3-658-41317-0_6

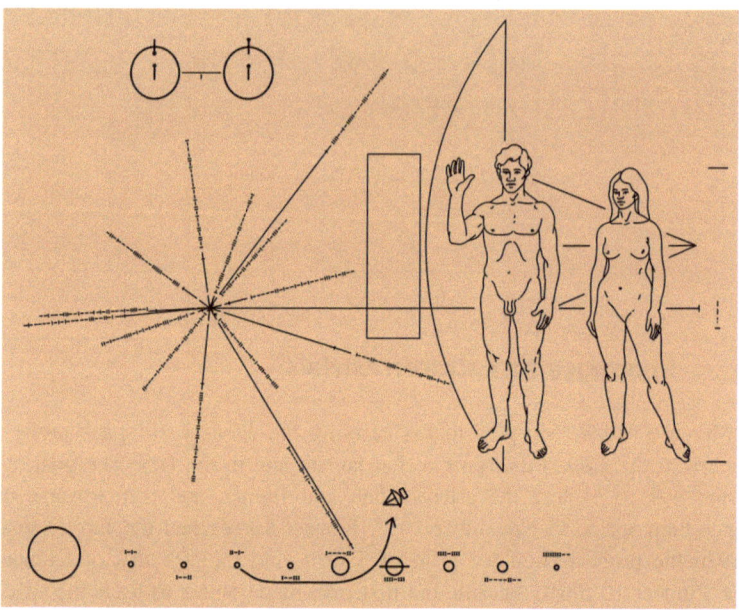

Fig. 6.1 Message to potential extraterrestrials: images on the Pioneer 10 plaque. *Source* public domain image[1]

The metal plaque attached to the Pioneer 10 spacecraft was the prelude to a whole series of projects aimed at transmitting human messages to possible extraterrestrial civilizations. They are referred to in scientific discussion as *Active SETI, METI* (Messaging to Extra-Terrestrial Intelligence) and occasionally *CETI* (Communication with Extra-Terrestrial Intelligence). The so-called *Arecibo Message*, a message to hypothetical extraterrestrials in the form of a radio wave signal sent on November 16, 1974 from a radio telescope near Arecibo (Puerto Rico) in the direction of the globular cluster M13 in the constellation Hercules, has achieved some notoriety. The content of the message, again conceived in large part by Frank Drake, was binary-coded information about human biochemistry and the position of the Earth in our solar system. The aim of the message broadcast, according to Drake, was above all to draw attention to humanity:

---

[1] Available online at: https://de.wikipedia.org/wiki/Pioneer-Plakette#/media/File:Pioneer_p laque.svg.

I thought that the message contained the information most relevant, interesting, and important to the creatures in space who might one day discover it. Yet I kept reassuring myself that the specific content was not actually the crucial aspect. Rather, the form of the message represented a message in itself that could be reinforced by periodic repetition. The transmitting power of the Arecibo transmitter was about half a million watts, but concentrated in a beam with an effective power of about 20 trillion watts. On its specific wavelength, the message would shine brighter than the sun. This, I believed, should be sufficient to attract attention. (Drake and Sobel 1994, p. 265)

The next message to hypothetical extraterrestrial intelligences came in 1977. In August and September, the two space probes *Voyager 1* and *Voyager 2 were* launched with the aim of exploring the outer planets of the solar system and interstellar space. As with the previous Pioneer mission, the Voyager probes contained messages to potential extraterrestrial recipients—but this time in the form of gold-plated copper plates, *the so-called Voyager Golden Records,* which contain both image and audio data. Among them are images of Earth, Mars and other planets of the solar system, pictures of human anatomy, DNA, images of different animal species, architecture and technology, but also of different human everyday situations. In addition, the data carriers contain a selection of music (from classical to rock 'n' roll) as well as greetings to the extraterrestrials in 55 different languages,[2] including that of the then UN Secretary-General Kurt Waldheim:

As the Secretary General of the United Nations, an organization of 147 member states who represent almost all of the human inhabitants of the planet Earth, I send greetings on behalf of the people of our planet. We step out of our solar system into the universe seeking only peace and friendship, to teach if we are called upon, to be taught if we are fortunate. We know full well that our planet and all its inhabitants are but a small part of the immense universe that surrounds us and it is with humility and hope that we take this step. (Teltsch 1977)

The 'material messages' of the Pioneer and Voyager probes, however, remained exceptions on the whole. In the following years, METI projects concentrated

---

[2] On a NASA homepage about the Voyager missions you can find an overview of the different image and audio files on the Voyager Golden Records: https://voyager.jpl.nasa.gov/.

mainly on the transmission of radio messages. In the following, we would like to cursorily present some projects of this kind:

- *Cosmic Call 1* (1999) and *Cosmic Call 2* (2003): In these privately funded projects, radio messages were sent from a radio antenna in Ukraine, to several stars near the Sun. The Russian radio astronomer Aleksandr Zaitsev was the consultant scientist for the projects (Chorost 2016; Zaitsev 2006).
- *Teen Age Message* (2001): In this project, also from Ukraine, a series of radio messages was sent to several sun-like stars, containing, among other things, music selected by Russian teenagers. This project was also accompanied scientifically by Aleksandr Zaitsev (Zaitsev 2008).
- *A Message from Earth* (2008): This project transmitted a powerful digital radio signal to the exoplanet *Gliese 581*.[3] The signal contains several hundred messages from people who had previously been selected within a social network. The signal will reach *Gliese 581 c* in 2029. This message was also transmitted from Ukraine (from the RT-70 radio telescope in Yevpatoriya). The scientific advisor was again Aleksandr Zaitsev (Moore 2008).
- *Hello from Earth* (2009): Similar to the *A Message from Earth* project, text messages (over 25,000) that could be sent in beforehand were transmitted into the Gliese 581 system. The transmission was made via a radio antenna of NASA's *Deep Space Network* in Australia.
- *Wow. Reply* (2012): On August 15, 2012, the Arecibo radio telescope sent an unofficial 'reply' to the Wow signal[4] in the direction from which that mysterious signal had come 35 years earlier. The content of the message was around 10,000 previously collected messages from the short message service *Twitter* (Gerritzen 2016, p. 221).

The most active advocate for METI at present is probably the US psychologist Douglas Vakoch, who in numerous publications promotes the idea of supplementing the passive search for extraterrestrial civilizations by actively establishing contact and sending strong radio signals into space for this purpose. Vakoch sees several advantages in this strategy compared to the usual passive search strategy of SETI:

---

[3] We had already devoted ourselves to this exoplanet in Sect. 3.3.
[4] See Sect. 4.2.

Complementing this existing stress on Passive SETI with an additional commitment to Active SETI, in which humankind transmits messages to other civilizations, would have several advantages, including (1) addressing the reality that regardless of whether older civilizations should be transmitting, they may not be transmitting; (2) placing the burden of decoding and interpreting messages on advanced extraterrestrials, which may facilitate mutual comprehension; and (3) signaling a move toward an intergenerational model of science with a long-term vision for benefiting other civilizations as well as future generations of humans. (Vakoch 2011, p. 476)

Vakoch especially deals with the question how a human message would have to be composed so that it could be decoded and understood by an extraterrestrial civilization with a high probability. In his view, the interstellar messages used so far are rather unsuitable for establishing communication with an extraterrestrial intelligence. He pleads for messages that are oriented towards basic mathematical statements (Marsiske 2007) and would like to send them into space as soon as possible so that future generations can hope for an extraterrestrial answer to our 'calling in the dark forest' as soon as possible:

First, if the goal of Active SETI is to initiate contact that leads to a response from extraterrestrials to future generations of humans, then the longer we wait to begin a serious transmission process, the longer future generations of humans will need to begin listening for replies that could answer the question of whether there are civilizations out there, ready to reply if only humankind takes the initiative to begin the conversation. (Vakoch 2011, p. 487)

6.2 Criticism and Potential Risks

From the beginning, the various human messages to potential extraterrestrial civilizations aroused not only enthusiasm, but sometimes also fierce criticism, both among the public and in scientific discourse. In the case of the material messages of the Pioneer and Voyager missions, this related primarily to the *content of* the messages. Criticism was voiced, among other things, that the messages were selected by a small (elite) group of people, who would thus have unjustifiably chosen themselves to be representatives of all of humanity. Furthermore, the messages would convey a distorted or incomplete picture of the realities of human life on earth, since topics such as war, poverty and hunger would not play a role. It is also highly questionable whether extraterrestrial intelligences would even be able to decipher and understand the very anthropocentric content of the messages (Sample 2018; Schulze-Makuch 2018).

However, the much greater part of the criticism of METI related and still relates to the radio messages sent into space. The most prominent critic, namely of Alexander Zaitsev's projects, was the recently deceased astrophysicist Stephen Hawking, who repeatedly warned strongly against sending further messages to potential extraterrestrials into space via radio waves. Hawking argued that we have no way of knowing whether potential alien civilizations are sympathetic or more hostile to us, so we should rather 'keep quiet' and not specifically call attention to ourselves. For him it is clear that contact with an extraterrestrial civilization could have drastic effects of a negative nature on humanity: "If aliens visit us, the outcome would be much as when Columbus landed in America, which didn't turn out well for the Native Americans" (quoted from Hickman 2010). Hawking is supported by a number of SETI scientists, who in 2015 wrote a joint statement expressing their displeasure with METI. In it, the signatories, who include Elon Musk, founder of the private space company SpaceX, warn of the potential consequences of radio messages sent into space and criticize previous METI projects for allowing individuals to make decisions that could affect all of humanity. The reaction of a potentially technologically advanced extraterrestrial civilization to our messages cannot be predicted, they say. Before any more messages are sent out, they say, there must first be deep debate and global consensus:

> We feel the decision whether or not to transmit must be based upon a worldwide consensus, and not a decision based upon the wishes of a few individuals with access to powerful communications equipment. We strongly encourage vigorous international debate by a broadly representative body prior to engaging further in this activity. [...] Intentionally signaling other civilizations in the Milky Way Galaxy raises concerns from all the people of Earth, about both the message and the consequences of contact. A worldwide scientific, political and humanitarian discussion must occur before any message is sent. (Azua-Bustos et al. 2015)

At its core, the criticism of METI involves two aspects: First, there is the question of the legitimacy with which previous METI projects have sent messages into space that, assuming potential extraterrestrial recipients, represent *all of humanity*. Secondly, there is a fundamental disagreement about whether it is wise, to take up the metaphor on which the title of this chapter is based once again, to call attention to oneself in a *dark forest*[5] by shouting, since one does not know whom or what one is attracting.

Assuming that the concrete reaction of possible extraterrestrial recipients of our interstellar messages also depends on their content, it seems sensible to think carefully beforehand about what information one reveals about oneself. In 2005, at a conference in San Marino, the Hungarian astronomer Iván Almár proposed a scale that could be used to quantify human messages sent into space. The decisive criteria of the *San Marino scale,* named after the conference venue, are the type and amount of information transmitted and the signal strength. The more information a message contains and the stronger the radio signal used to send it, the higher its value on the scale—and the higher the potential consequences (both positive and negative) of an extraterrestrial 'response'. The scale ranges from 1 (insignificant) to 10 (extraordinary). Thus, according to the logic of the San Marino scale, if humanity transmits strong signals with a lot of information about itself, the risk of unwanted reactions from extraterrestrial intelligences increases (Almár & Shuch 2007).

Admittedly, the San Marino scale—as well as the entire debate on METI—is highly influenced by anthropocentric presuppositions. Ultimately, it is impossible to predict in any way whether and how any extraterrestrial civilizations would react to human radio signals—not even whether they would be able to understand them at all. Nevertheless, we think the objections to METI are essentially valid. The question of whether we as a human species send messages into space (and if so, with what content) should not be decided by individuals. Since quite a number of nations have radio telescopes that are capable of sending a message with high transmission power into the vastness of space, a binding regulation under

---

[5] We take this metaphor from the novel of the same name by the Chinese SF author Cixin Liu (Liu 2018). In the context of a lengthy passage, he describes the protagonist's deliberations that eventually lead to averting an alien invasion with the threat of shouting out the position of the invaders' home planet into space, as it were. The captivating thing about this resolution of the Fermi Paradox is that the Dark Forest thesis holds regardless of whether alien civilizations are judged to be friendly, ethically neutral, or more malevolent by a species' particular standards. Before the imaginable first contact, such a classification is in any case completely uncertain—and also at a later point in time, the limits of alien understanding will make it extraordinarily difficult to make an appropriate classification here (mutually!). A mistake in the assessment could lead to the complete destruction of one's own culture. Therefore, every civilization seems to be held to hide its own existence from the others as well as possible at the price of its survival. For exosociology, this explanatory model would have the unpleasant consequence that interstellar cultural contact would become very unlikely, thus raising the question of the raison d'être of this subdiscipline. Fortunately, there are (as we pointed out in the previous chapter) a number of other, much less radical solution models for the Fermi paradox.

international law in this area seems urgently necessary. In our opinion, it would be optimal if the decision on METI experiments (and likewise on the question of the response to a received extraterrestrial signal) were made by a body, such as a kind of 'foreign committee' of the United Nations. This would prevent individual states or even groups of researchers from making decisions that could have momentous consequences for the whole of humanity. The problem also seems to us to be too serious to be left to purely scientific expert bodies. In the present state of the human 'community', however, it seems questionable to us whether a corresponding binding regulation can be achieved in the foreseeable future. From an exosociological point of view, one aspect of the controversy about METI also seems to us to be particularly interesting: the possible courses of action of *hypothetical* extraterrestrials serve as evaluation criteria for *real* human actions–both for METI proponents and critics. The prerequisite for the argumentation of both sides is, of course, that we are not alone in the dark forest.

References

Almár, Iván & Paul H. Shuch. 2007. The San Marino Scale: A New Analytical Tool for Assessing Transmission Risk. *Acta Astronautica* 60 (1): 57–59.

Azua-Bustos, Armando et al. 2015. Regarding Messaging to Extraterrestrial Intelligence (METI)/Active Searchers for Extraterrestrial Intelligence (Active SETI). https://setiathome.berkeley.edu/meti_statement_0.html.

Chorost, Michael. 2016. How a Couple of Guys Built the Most Ambitious Alien Outreach Project Ever. Smithsonian.com am 26. September 2016. https://www.smithsonianmag.com/science-nature/how-couple-guys-built-most-ambitious-alien-outreach-project-ever-180960473/?no-ist.

Drake, Frank, und Dava Sobel. 1994. *Signale von anderen Welten. Die wissenschaftliche Suche nach außerirdischer Intelligenz.* München: Droemer.

Gerritzen, Daniel. 2016. *Erstkontakt. Warum wir uns auf Außerirdische vorbereiten müssen.* Stuttgart: Kosmos.

Hickman, Leo. 2010. Stephen Hawking Takes a Hard Line on Aliens. *The Guardian*, April 26, 2010. https://www.theguardian.com/commentisfree/2010/apr/26/stephen-hawking-issues-warning-on-aliens.

Liu, Cixin. 2018. *Der dunkle Wald*. München: Heyne.

Marsiske, Hans-Arthur. 2007. Welche Sprache sprechen Außerirdische? Welt Online vom 09. Dezember 2007. https://www.welt.de/wissenschaft/article1439767/Welche-Sprache-sprechen-Ausserirdische.html.

Moore, Matthew. 2008. Messages from Earth Sent to Distant Planet by Bebo. *The Telegraph*, October 9, 2008. https://www.telegraph.co.uk/news/newstopics/howaboutthat/3166709/Messages-from-Earth-sent-to-distant-planet-by-Bebo.html.

Sample, Ian. 2018. Nasa's Golden Record May Baffle Alien Life, Say Researchers. *The Guardian*, May 26., 2018. https://www.theguardian.com/science/2018/may/26/nasas-gol den-record-may-baffle-alien-life-say-researchers.

Schulze-Makuch, Dirk. 2018. How to Communicate with Aliens. Some Interesting Ideas Bounced Around at a Recent Workshop. Air & Space Smithsonian. https://www.airspa cemag.com/daily-planet/how-communicate-aliens-180969211/.

Teltsch, Kathleen. 1977. U. N. Sending Messages Aboard Voyager Craft for Beings in Space. *New York Times*, June 3, 1977. https://www.nytimes.com/1977/06/03/archives/un-sen ding-messages-aboard-voyager-craft-for-beings-in-space.html.

Vakoch, Douglas. 2011. Asymmetry in Active SETI: A Case for Transmissions from Earth. *Acta Astronautica* 68: 476–488.

Zaitsev, Aleksandr L. 2006. Messaging to Extra-Terrestrial Intelligence. arXiv:physics/061 0031. https://arxiv.org/ftp/physics/papers/0610/0610031.pdf.

Zaitsev, Aleksandr L. 2008. The First Musical Interstellar Radio Message. *Journal of Communications Technology and Electronics* 53 (9): 1107–1113.

Methodological Considerations for the Scenario Analysis of First Contact

As long as humanity does not yet possess empirical knowledge about an extraterrestrial intelligence that makes sociological explorations of alien society meaningful and necessary, probably the most important task of exosociology (already the progenitor of the subdiscipline, Jan H. Mejer, had seen it this way more than thirty years ago) is the prognostic assessment of the terrestrial consequences of humanity's confrontation with the *certain knowledge* that, as an intelligent species, it is not alone in the universe. Before we ask in the following chapter what the possible and probable consequences of such an event might be, we must first clarify how it is possible at all to make scientific statements about an event that—at least according to the firm conviction of most scientists working in this field—has not yet occurred.[1] This is obviously a task for scientific futurology or futurology,[2] which for many decades has developed a whole range of methods for obtaining at least more or less uncertain knowledge about what one certainly cannot know, namely what the future will bring. Of the various methods of futurology, in order to answer the question of the possible consequences of such contact for human civilisation, we want to make use of that method which, on the one hand, makes the fewest demands with regard to the data situation for extrapolating previous developments into a still uncertain future and, at the same time, is capable of comparatively investigating a whole series of possible futures: *scenario analysis*.

---

[1] The different theses of the so-called pre-astronautics will be discussed in Chap. 11.

[2] For a culturally critical overview of the history of futurology and its methods, see Uerz (2006, pp. 257–319).

7.1   The Method of Scenario Analysis

Scenario analysis has been one of the most important methods of scientific foresight for decades.[3] Usually, the aim is to extrapolate political or economic developments for which qualitative and/or quantitative data are available for the past into the near or distant future (not necessarily meant mathematically here). The special feature of this method is to sketch a whole space of possible futures by distinguishing more or less probable,[4] as it were ideal–typical developments in the field under investigation (Zürni 2004, pp. 14–15 and pp. 132–135; Fink and Siebe 2006, pp. 15–16; Berghold 2011, pp. 27–29). The individual, analytically distinguishable ideal–typical developments are thereby referred to as 'scenarios' (which gave this method its name). Depending on the area of investigation, the scenarios can be sometimes more, sometimes less complex; the complexity is determined on the one hand by the number of influencing factors taken into account and on the other hand by the complexity of the descriptions of possible futures—for example with regard to the consequential dimensions examined (Berghold 2011, p. 57). In the simple case, a more formal distinction is made between three scenarios: a worst-case scenario, a best-case scenario and an expectable ('average') scenario.[5] Alternatively, two key factors, each with dichotomous trends, can be set against each other, resulting in a 'four-field table' whose fields each correspond to a scenario. More complex analyses with a large number of different scenarios are also conceivable, although they are much more difficult to handle in practice. The method-oriented literature therefore recommends limiting the scenarios analysed in a process to a number of three to six (Zürni 2004, p. 146).

Since the scenarios are described with quantitative, qualitative or both types of data, this method can be used to determine future scenarios in very different fields of action and to forecast their effects—for example, the development of crude oil prices, changes in international trade policy or the rise and fall of political movements.[6] In these 'normal cases', a set of factors is always available of which

---

[3] An introduction to scenario analysis is provided by Fink and Siebe (2006, pp. 15–79); the method is presented in detail in the volumes by Zürni (2004) and—from a strictly economic perspective—by Berghold (2011). In our own scenario analysis in the following main chapter of this volume, we have essentially drawn on these three volumes.

[4] In contrast to many other methods of futurology, the scenario technique usually avoids attempts to determine the probability of occurrence of individual scenarios more precisely.

[5] Berghold (2011, pp. 30–31) distinguishes somewhat generally trend scenarios from extreme scenarios.

[6] Fink and Siebe (2006, pp. 15–79) present a whole series of example analyses in detail.

it is known, on the basis of past developments, what influence they have had on the political, economic or even ecological field under investigation (for example, the connection between commodity prices and the threat of war). The method is primarily regarded as 'inductive'[7] because the future scenarios are usually (but not always) developed from *empirical* data of the past and present—this question, however, seems to us to be of little fundamental importance, rather secondary, as long as the extraction of the initial data for the analysis is transparent.

The various scenarios result from the synopsis or comparison of the various development possibilities of those key factors (Zürni 2004, pp. 222–225; Berghold 2011, pp. 42–44). The more key factors are identified, the more complex the scenario analysis becomes—and the more possible baseline scenarios (sometimes also called 'raw scenarios' in the literature) need to be examined. With a larger number of key factors, the field of conceivable scenarios quickly becomes unmanageably large, so that the analysis must settle on a few main scenarios, which are then analysed in more detail. In practice, the primary problem with this method is the rationale for selecting the main scenarios to be examined or constructed in more detail.[8] A similar problem arises when determining the consequential dimensions to be examined: i.e., for example, the economic or political fields on which the influencing factors have a direct or even indirect effect. If too many dimensions are included in the analysis, the picture becomes overly complex—Berghold (2011, p. 44) rightly points out that the analysis effort increases exponentially, with the number of influencing factors considered, due to the interactions that usually exist. In addition, with a large number of factors examined, the individual scenarios become increasingly difficult to distinguish and ultimately also threaten to lose their informative value as, as it were, ideal–typical future options.

The literature contains a whole series of proposals for the concrete procedure of a scenario analysis (see Zürni 2004, pp. 167–176). A comparatively simple and therefore easily manageable process model has been presented by Fink and Siebe (2006, pp. 37–49); they propose four work steps ('phases') for carrying out scenario analysis: (1) selection of key factors, for each of which there are different development alternatives, (2) formulation of future projections based on the possible future states of these factors, (3) integration of the various projections into a limited number of distinguishable scenarios, and (4) detailed, possibly 'pictorial' description of these scenarios (see also Berghold 2011, pp. 38–55).

---

[7] The differences between inductive and deductive procedures within the spectrum of this method are discussed in detail by Zürni (2004, pp. 167–176).

[8] "There is no clear procedure for selecting the most relevant scenarios" (Zürni 2004, p. 225).

In the case of first contact of mankind with an extraterrestrial civilization that interests us, we will apply the method of scenario analysis in an unusual way to a so-called *wild card event*. Such events are characterized by the fact that although the probability of their occurrence is low, if they did occur there would be significant impacts that would massively affect individual or a large number of subsystems of society. Angela and Heinz Steinmüller (2004) have systematically examined the significance of such events.[9] *Wild cards* are usually single events, but they can also be the culmination of longer-term processes—for example, a terrorist attack, a stock market crash or a natural event such as a volcanic eruption, which are announced by various 'omens'. Some of these events do not come 'out of the blue', but can be recognised by what futurologists call 'weak signals' a limited time before they occur. Other events, however, (such as the impact of a previously undiscovered asteroid on Earth) occur without any warning. What these events have in common (this is the first problem from a methodological point of view) is that the probability of their occurrence cannot be estimated numerically:

> With wild cards, however, our prior knowledge is very limited. The evaluation of probabilities is correspondingly poor. Even the circumstances under which a disruptive event is possible in principle can be disputed. [...] When it comes to estimating the probability of wild cards, all mathematical approaches usually fail; at best, arguments pro and con and analogies help a little. Wild cards are unlikely, but we do not know how unlikely they are. (Steinmüller and Steinmüller 2004, pp. 19–20; Berghold 2011, p. 30)

A second problem is that the consequences of such abrupt and rare events are particularly difficult to predict. This is particularly because there are almost always secondary and tertiary consequences (and so on) in addition to the immediate consequences of the event. In a sentence, "The impact of wild cards is so great because they break out of the usual frame of reference, shake it up, challenge normality, subvert the thought patterns with which we construct the world" (Steinmüller and Steinmüller 2004, p. 22). It should be added that the effect of wild cards does not extend to the future alone; with their occurrence they also change the interpretation, perhaps the perception, of the past. "For this is now

---

[9] In classical scenario analysis, little space is usually devoted to such events under the heading of 'disruptive event analysis' (Berghold 2011, pp. 51–52). This is primarily due to the fact that analyses conducted in the economic context are based on an explicit or implicit continuity paradigm—even extreme scenarios often reflect a 'business as usual' mentality. Against this background, the deeper meaning of the terminology *'disruptive event'* also unfolds.

thought in terms of the new patterns, events and developments are re-sorted and evaluated in terms of the changed end" (Steinmüller and Steinmüller 2004, p. 23). In general, it can be said that the social consequences of a wild card event depend, on the one hand, on the parameters of the event itself (such as the magnitude and epicentre of an earthquake) and, on the other, on how society deals with it. In the latter case, in turn, an analytical distinction must be made between societal action before the event occurs (precaution) and after it has occurred (reaction). Whether precautions are taken, in the sense of preliminary measures to prevent or mitigate certain consequences of the event, depends less on its *scientific expectability* (the assessment of the probability of occurrence by experts) than on its *social expectation* (convictions of the population plus recognition of the need for action by state authorities).[10]

In the context of scenario analyses, wild card events are usually recorded in the form of a so-called *disruptive event analysis*. Their investigation begins with a number of questions (according to Steinmüller and Steinmüller 2004, p. 31): What does the disruptive event consist of? What is the probability of occurrence (if it can be estimated at all)? Are there preconditions for the occurrence of the event? Which societal subsystems and regions of the globe are primarily affected? What are the direct and indirect effects on society as a whole? With what temporal, spatial and systemic dynamics will the impacts unfold?

In addition to the analytical proposals of Steinmüller and Steinmüller, it seems useful to us to distinguish between two ideal types in the analysis of wild card events: (1) Events, such as volcanic eruptions, epidemics, economic crises, terrorist attacks or wars, which have already taken place in one form or another and whose effects on the societies concerned are at least generally known. (2) Events of a kind that have not yet occurred in recorded human history. Classic examples

---

[10] The fact that there is no compelling connection between the two can be shown empirically in a number of cases. A striking current example of this is climate change and the collapse of the North Atlantic Current that it is likely to trigger. The 'breakup of the Gulf Stream' is now expected by many climate and ocean scientists to occur in the next few decades, although the timing of its occurrence is currently difficult to predict. The effects of this event would be ecologically, economically, socially and politically devastating for Northern and Central Europe. Despite this (or perhaps because of it), the very real danger is ignored by political decision-makers as well as most inhabitants of the affected regions—perhaps also due to a media-generated reality shift that increasingly degenerates climate change from a 'real social problem' to science fiction.

of this are the dropping of the atomic bomb on Hiroshima at the end of the Second World War or the Chernobyl disaster in 1986.[11] The central characteristic of these events, which were still *hypothetical* before they first occurred, is that their consequences are virtually 'unforeseeable' in a metaphorical sense, which means that forecasting their effects is extremely difficult.

In our opinion, hypothetical events in this sense also include the first contact of mankind with an extraterrestrial civilization,[12] about which at least the relevant sciences agree that it has not yet occurred in the historically handed down history of mankind. For such an event, what Steinmüller and Steinmüller (2004) wrote about the consequences of some global[13] wild cards would certainly apply: Their effects "sometimes assume an almost fatal dimension" (p. 38), they have "the power to trigger shock waves of change" (p. 13). First contact with an extraterrestrial civilization, all experts ultimately agree, would be one of the most drastic events in human history.

Because of the uniqueness of hypothetical events, it is methodologically appropriate when forecasting their effects to additionally resort to the technique of creating narrative scenarios (Fink and Siebe 2006, pp. 56–64), since this procedure is less strict in dealing with key factors. The focus here, as the name suggests, is on the densest possible narration of individual scenarios—which can, but does not have to, take the form of fictional texts. In creating the first contact scenarios described in detail in the next main chapter, we also made use of narrative elements, in addition to the classic scenario technique.

---

[11] It is no coincidence that the examples mentioned are both related to the use of nuclear power—events such as those in Hiroshima, Chernobyl or Fukushima are only possible in the history of mankind after the corresponding nuclear technologies have been developed.

[12] Steinmüller and Steinmüller (2004, pp. 86–87) also devote a short chapter to this event—although they limit themselves to the case of the reception of an extraterrestrial radio signal, to which they attribute rather minor cultural effects compared to other wild card events.

[13] First contact with an extraterrestrial civilization undoubtedly represents a global event with corresponding effects in the information-networked world. Such contact in historical times, namely in human prehistory or early history, on the other hand, could well be conceived as a local event, affecting only the population of a narrowly defined region.

7.2 Parameters of the Scenario Analysis of the First Contact

The analysis of a wild card event should (see Steinmüller and Steinmüller 2004, pp. 31–32) on the one hand avoid all superfluous or unfounded presuppositions, and on the other hand reveal the necessary preconditions of its analysis. The scenario analysis we conducted on first contact has four basic *preconditions* (Schetsche 2008, 228–230):

1. *There is the possibility of the coexistence of civilizations in the cosmos that are willing to communicate:* A scientific preoccupation with the possible effects of first contact only makes sense if there is at least a certain probability that humanity is not alone in the universe at the time of its existence as a technical civilization. As with the SETI programmes (Engelbrecht 2008), it must be assumed that there *could be* extraterrestrial civilisations in the vastness of space with which contact—by whatever means—is in principle possible. How this could happen and what the consequences would be in each case is the subject of the scenario analysis.

2. *The extraterrestrials must be recognizable as such:* A cultural contact[14] only takes place in our perception if we are able to identify the others as intelligent non-terrestrial entities. This is not a matter of course; rather, situations can be imagined (science fiction knows corresponding thought experiments) in which intelligent beings do not recognize each other as such even in direct contact—for example, because their perceptual spaces are not compatible or because their perceptions of time diverge too much (Bach 2004; Schetsche 2004).

3. *There is an inescapable facticity of not knowing:* before the actual contact we cannot know anything about the physical and psychological constitution, about possible social structures or even the interests and motives of the strangers. This forces us to strictly ignore all 'qualities' of the others when considering the consequences of a first contact. At the present time, we can only think about the consequences of a first contact for the Earth on the basis of our knowledge of our human thought structures and (collective) behaviour patterns.

---

[14] Here understood in the most abstract sense: Mankind for the first time in its history acquires certain knowledge about the existence of an extraterrestrial civilization—whether this leads to mutual exchange of information or interaction in the true sense is quite another question, depending primarily on the nature of the contact (this we examine in the following Chap. 8).

4. *Anthropocentric presuppositions should be minimized.* Since we can literally know nothing about the aliens before contact (see 3. Precondition), we have to say goodbye in particular to all the anthropocentric prejudices by means of which we regularly 'humanize' aliens in our thinking—in science fiction and in science. This includes, (we have already addressed this earlier in Chap. 4) in particular, the assumption that extraterrestrials, due to the 'insurmountably' of large distances in the cosmos, are always only able to 'speak' to us from a great distance (Michaud 2007a, p. 123). The notion that a direct physical encounter is and will remain impossible only seems conclusive as long as a number of highly dubious presuppositions are made: human-like travel technology and temporality of the extraterrestrials, subject-oriented travel planning, or even the 'biological quality' of potential visitors (Kuiper and Morris 1977; Michaud 2007b, p. 2). Another presupposition to be avoided is the idea (still widespread among the public) that aliens would confront us as a biological species to which the same rules (such as those of evolutionary biology) apply as to ourselves. To us, on the other hand, it seems at least as likely (if not more likely[15]) that those aliens are a post-biological secondary civilization of very advanced 'machine beings'. In this respect, too, a scenario analysis should make no presuppositions about the 'nature' of those Others—the considerations should be entirely independent of what kind of aliens we encounter.

Based on these preliminary assumptions, it is already clear that scenario analysis of the consequences of first contact for human civilization as a whole cannot, as is classically the case with this method, construct different scenarios with the same initial conditions, in which different developments of several key factors lead to readily distinguishable 'end states' after a certain forecast period. Too

---

[15] How one evaluates this probability depends in particular on one's assessment of how long the average time period is between the development of technologies suitable for interstellar communication and the replacement of the primary, biologically evolved intelligent species by a secondary machine civilization. This is difficult to estimate because we have exactly zero example cases here at the moment. Mankind has only been engaged in space travel for a few decades (though not even an interstellar one), but is already planning technological projects that could lead to its replacement by a machine intelligence in the medium term. For example, futurologist Bostrom (2014) anticipates the emergence (more correctly: development) of a machine superintelligence that will—within this century—take over power on Earth from humanity. If this prediction is even tendentially accurate, it could mean that the window of opportunity for a biologically evolved species to conduct systematic space exploration before being replaced is at best a few centuries long, or rather short. (We return to this question in Chap. 10).

few parameters are available for such an approach to vary in a systematic and controlled way. Perhaps the most important cause of this is the lack of any information about the courses of action available to the aliens, about their motives and interests. We are repeating ourselves here, but this cannot be stressed often enough: Since we still know literally *nothing* about the existence, technical possibilities, interest positions, etc. of extraterrestrials, the terrestrial consequences of this contact can only be predicted on the basis of parameters that are *independent of the characteristics, etc. of the aliens themselves* (Schetsche 2003, p. 26).

Scenario analyses that use imaginary motives and interests of the extraterrestrials (e.g. in the sense of peaceful cooperation vs. conquest of the Earth[16]) as key factors for assessing the consequences of the event are, in our opinion, misleading because they are anthropocentric in two senses: On the one hand, they assume that the extraterrestrials have similar motivations as we know them from humans, or more correctly from the terrestrial nation states of the last centuries; on the other hand, it is also implicitly assumed that from the observable actions of the extraterrestrials in the case of contact (see 'encounter scenario' in the following chapter), their interests can be concluded without further ado. If we follow the *theory of the maximum alien* (Schetsche et al. 2009), both assumptions are invalid. A scenario analysis of the consequences of first contact for humanity that uses such or similar parameters would, in our view, be purely speculative. This does not mean that such considerations should not be made—they have had a firm place in science fiction for decades anyway (Engelbrecht 2008; Hurst 2008). In a *scientific* prognosis of the impacts, on the other hand, primarily those dimensions/parameters should be used and investigated that relate to human action, in particular to the development as well as the reactions of terrestrial societies to extraordinary 'disruptive events'.[17]

---

[16] Especially the sweeping ethical distinction between 'good' and 'evil' intentions of extraterrestrials is not very purposeful, since it not only assumes that extraterrestrials have an ethical evaluation system, but also tries to orient it to human standards. Distinctions of this kind, we are convinced, merely generate a multitude of analytical misunderstandings (not to say avoidable errors).

[17] Not to be misunderstood: Of course, it makes a difference for the terrestrial reactions whether extraterrestrial spacecraft in the solar system react passively-reluctantly (by human standards) or, from our point of view, undertake acts that can only be interpreted as warlike—the range of externally observable actions of the spacecraft and, even more so, of the subsequent human interpretations (and speculations) is so great that almost any number of individual scenarios could be designed here. Precisely in order to minimize the highly speculative moments, we have concentrated in the following chapter on a simple scenario that is rather 'action-poor' on the part of the aliens; this makes it possible to focus on the reactions of human society(ies) in the analysis.

We will demonstrate how a forecast is possible under this basic assumption in the following Chap. 8. There you will find the results of our scenario analysis based on a series of initial scenarios that differ in a single, but absolutely central parameter: *the form in which the initial contact takes place.* And this form (this is a fifth, equally structural and methodological presupposition of our analysis) can be determined on the basis of two distance parameters at the contact event: the spatial and the temporal distance between us and 'the others'. The following graph maps these two distance parameters on its two axes, creating a prognostic space of the technical-structural forms of first contact (see Fig. 7.1).

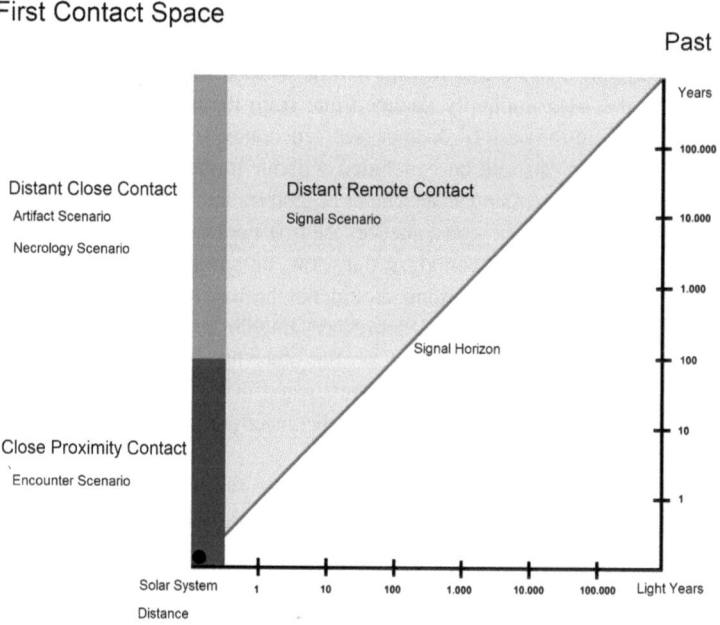

Fig. 7.1 Spatial and temporal distance as primary parameters of the scenario analysis. *Source* own representation

The diagram shows that three basic scenarios (or basic situations) can be distinguished for forecasting the earthly consequences of the wild card event of *first contact*, each of which would have quite different consequences:

I. *The distant remote contact*: This basic scenario has dominated the scientific debate about the possibility of human-alien contact for many decades—it reflects the basic assumption of traditional SETI research, according to which that first contact will take the *form of* the reception of an extraterrestrial signal reaching us by electromagnetic means from a distance of many (hundreds or even thousands) of light years This basic scenario has two crucial consequences: First, direct (physical) contact between humans and extraterrestrials remains impossible for the unforeseeable future (perhaps even forever). Secondly, the basic communicative situation is a monological one due to the long signal travel times—a dialogue cannot take place within the framework of a human time horizon, since every question as well as every answer travels for centuries or even millennia before it reaches the counterpart. Based on these basic assumptions, we have developed our *signal scenario* for the following chapter.

II. *The distant close contact*: As noted earlier, we understand 'distance' in a dual sense: spatial and temporal. These dimensions behave asymmetrically to each other: due to the limitedness of the speeds possible in the universe (of signals as well as of objects),[18] spatially close 'encounters' are possible over long periods of time (for example, with ancient extraterrestrial artifacts), but not temporally close ones, which refer to large spatial distances—a signal from a planet 1,000 years away is of necessity 1,000 years away, so always refers to a correspondingly large time horizon. Spatially near and temporally distant[19] confrontations with an extraterrestrial intelligence are represented in the form of two scenarios, the *artifact scenario* and the *necrology scenario*.

---

[18] We assume here and in the rest of the volume the current human knowledge of the nature of the cosmos and the validity of the laws of physics (such as to that velocity maximum). It makes little sense, in our view, in the context of this volume, to speculate about whether there is any 'alien physics' that makes faster signal travel times or even faster-than-light spaceship propulsion possible. As a scientific enterprise, exosociology must be guided by the currently valid descriptions of its neighbouring scientific disciplines. We leave techniques like the "warp drive" (from the 'Star Trek universe') to science fiction.

[19] For the determination of temporal distance, we take the three-generation rule from Maurice Halbwachs theory of collective memory (1967) as a basis: everything that lies further back than 100 years is (inter)subjectively 'distant' for living people.

III. *Close proximity contact*: In this basic scenario, small spatial distance and small temporal distance coincide. Spatially, we focus here on 'our' solar system with its known planets and dwarf planets, i.e. a space around our Sun measuring considerably less than one light year. This distance limit is more a cultural than an astrophysical limit[20]: The space 'occupied' by the planets and dwarf planets *named* by humans has become so anchored in the (scientific) everyday consciousness through media reporting on interplanetary research that, in our estimation, a scientific-technical limit has also become a psychological limit ... or is likely to become one in the coming decades. This is not least due to the fact that this is the space that we, as acting human beings, are able to reach with the artificial objects we have created, the space probes, and in the near future perhaps even with spaceships.[21] According to our analyses, this basic spatiotemporal configuration manifests itself in an *encounter scenario*.

Here is an overview of the four scenarios we distinguished in our analysis (details in the following chapter):

1. *The signal scenario*: we receive an (electromagnetic) signal from the vastness of space, whose source is far beyond the borders of our solar system, but which is undoubtedly of artificial origin.
2. *The artifact scenario*: we discover on earth or somewhere in 'our' solar system an artificial object that is certainly not of earthly origin and that has been there for at least a hundred years.
3. *The necrology scenario*: we discover on earth or in the solar system the ancient remains of an extraterrestrial being or a highly complex technical entity that is no longer capable of action.
4. *The encounter scenario*: an alien spacecraft capable of interaction arrives in the solar system, with which direct communication seems possible—regardless of

---

[20] Astrophysically, our solar system extends much further—materially interpreted at least to the outer boundary of Oort's Cloud, at a distance of perhaps 1.5 light-years from the Sun. In our impression, however, the smaller and larger ice bodies that can be found there are virtually absent from people's everyday consciousness, so that this space must be assigned the (cognitive and emotional) status of 'far out there'—which in the current early space age also means: far beyond our human reach.

[21] From a social-psychological point of view, every landing of a space probe on a planet is ultimately also a kind of taking possession—always for 'mankind', but in individual cases quite deliberately on behalf of individual nation states (Schetsche 2005b).

whether the missile is AI-controlled or has an alien biological intelligence on board (which, from a human point of view, should not be so easy to distinguish anyway).

The following chapter will take a closer look at the respective scenarios and their multiple effects on various social subsystems. Before doing so, however, we will briefly discuss the factors or scientific arguments that can be used to assess the consequences of the different types of first contact.

7.3 Prognostic Strategies and Problem Situations

In general, it has to be said (we already briefly pointed out) that *the quality of the prognosis of the consequences of wild card events is limited by the question of how often a corresponding event has already taken place*—i.e., how many corresponding historical cases are available for comparison purposes (and of course: how well they are documented). This is the reason why hypothetical events, such as first contact with extraterrestrials, are a *problematic case* from a prognostic point of view. How is it at all possible here to come to even reasonably certain conclusions about the impact of this very special event? In our opinion, four prognosis strategies are conceivable with which it should be possible, at least in principle, to predict the actually unpredictable: at least to some extent (see Table 7.1).

Strategy 1: The transfer of empirical results from other fields of research. Here, for example, one could take (as a starting point) findings from the sociology of crime, according to which people's sense of threat is all the stronger the more the encounter with a potentially dangerous counterpart is located in their own social living space. The feeling of unease is strongest when we are confronted with what we perceive as a threat within our 'own four walls'. In order to use such an argument in the field we are interested in, a rather broad (possibly exaggerated) analogy is required: The earth and its immediate surroundings are equated for the entirety of humanity with the dwelling of the individual citizen of the earth in terms of mass psychology. From this, one can conclude: the appearance of the aliens on Earth itself would then be the worst-case scenario, psychologically speaking. In other words: What we know of the fear of crime and the reactions guiding it should also apply to the confrontation with the 'maximum alien' (Schetsche 2004). This strategy has the advantage that a whole range of empirical findings from a wide variety of research fields is available for almost every conceivable case—but it is precisely this that leads to the difficult question

Table 7.1   First contact consequences—forecasting strategies

Forecasting strategy	Research logic, basic method	Benefits	Problem areas
1. Transfer of empirical findings from other research fields	Analogy	Complexity of the available data	Doubts about comparability; problem of generalisation
2. Evaluation of 'natural experiments	Social experiment	Reconstruction of real behaviour	Dependence on life-world data
3 Historical parallels	Historical analogy	Extensive historical material, if applicable	Variety of parameters; limits of transferability
4. Theory-based extrapolation	Deduction	Orientation on proven theoretical slides	Lack of empirical data; problems of applicability

Source Schetsche 2005a, slightly revised

of which analogies are permissible and which are not (Denning 2013). If we want to use such a strategy, good reasons would have to be given in each case as to why precisely the selected external findings are meaningful for the case under investigation.

2nd strategy: The evaluation of 'natural experiments'. Here one could cite, for example, the broadcast of the radio play *War of the Worlds in* 1938 (we will return to this in detail in Chap. 8 below). From the reactions of the population to a media event mistaken for real, one could infer the consequences of a real first contact. The advantage of this strategy is that it works with real observational data, which (at least in cases as spectacular as the one described here) have already been systematically reconstructed in scientific works (see for example Cantril 1940). So, in principle, we could be dealing with well-established empirical material. However, the number of corresponding natural experiments is limited—even more serious, however, seems to be the fact that the data collected at that time are today regarded very critically by many researchers (Pooley 2013); this method is also burdened with many uncertainties.

3rd strategy: The search for historical parallels. Another strategy could start from historical examples of intercultural contacts on earth. The methodological thesis here is that the experience of asymmetrical cultural contacts on Earth (see Bitterli 1986: passim) can also be applied to humanity's confrontation with an extraterrestrial civilization. Accordingly, it is asked what we can learn from the history of terrestrial civilizations with regard to the medium- and long-term consequences of the discovery of Earth by extraterrestrials. The advantage of such a strategy is that there is extensive data available on corresponding asymmetric cultural contacts on Earth. The disadvantages lie in the complexity of each individual case—and in the question, already left open in the first strategy, of the transferability of the results. In particular, it is necessary to consider which parameters were decisive for the historical development and which of them are also likely to be relevant for the case of interstellar cultural contact that is of interest (Dick 2013, 2014; Lowric 2013).

4th strategy: The extrapolative application of theoretical considerations: The starting point here could be, for example, theories and findings of social psychology (for example, stereotype research). From them, one could conclude, for example, that we would judge the aliens in the case of direct visual contact primarily on the basis of their appearance. That is, from associations of the external appearance with terrestrial beings known to us would be concluded on modes of perception, character traits, motives, etc. of the aliens. In this respect, one could (following a formulation by Heinrich Popitz) almost speak of a "preventive effect of not knowing": The less we know about the physical appearance of the aliens, the less image-bound stereotypes or even inherited behavioural schemata will influence our actions. Knowledge of the aliens' 'appearance' will therefore not lead us to understand them better, but merely to *misunderstand* them more quickly (Harrison 1997, p. 198; Harrison and Johnson 2002, pp. 103–104; Michaud 1999, pp. 266–267 This application is an extrapolation because the theories addressed (here social psychology) are projected onto an empirical field for which they were not actually intended: Behaviour towards the socially or culturally foreign is used to infer behaviour towards the maximally foreign. The advantage of such procedures is obvious: we can start from empirically well-documented and widely accepted theoretical foils. However, we have to extend their scope mentally in order to be able to transfer them to the case of first contact with extraterrestrials, which has not yet occurred. Probably here, too, many problems lie in the details of such an extension.

The listing of these four basic forecasting strategies should also have made it clear how great are the difficulties we face in predicting the outcome of a hypothetical event such as that of first contact. The caution with which scenario analysis must be viewed in the following chapter must be correspondingly great. Certain knowledge of what will happen when mankind confronts an extraterrestrial intelligence for the first time, cannot be reliably obtained before this event occurs. In this respect, this case is initially no different from other Wild Card events. However, due to the complete lack of previous events of the same type, the uncertainty of prognostic knowledge in this case is even much greater than in the case of volcanic eruptions and disease epidemics, wars between nation-states, or even major terrorist attacks, all of which have occurred several times in the surviving history of mankind and for which a wide variety of data are available to estimate the consequences of future events of the same type. Nevertheless, we maintain, the global consequences of first contact are not entirely unpredictable and, accordingly, are not entirely unpredictable. We just have to keep in mind that the space of uncertainty is particularly large for this very special prediction.

References

Bach, Joscha. 2004. Gespräch mit einer Künstlichen Intelligenz – Voraussetzungen der Kommunikation zwischen intelligenten Systemen. In *Der maximal Fremde. Begegnungen mit dem Nichtmenschlichen und die Grenzen des Verstehens,* Hrsg. Michael Schetsche, 43–56. Würzburg: Ergon.

Berghold, Christina. 2011. *Die Szenario-Technik. Leitfaden zur strategischen Planung mit Szenarien vor dem* Hintergrund *einer dynamischen Umwelt.* Göttingen: Optimus.

Bitterli, Urs. 1986. *Alte Welt – neue Welt. Formen des europäisch-überseeischen Kulturkontaktes vom 15. bis zum 18. Jahrhundert.* München: Beck.

Bostrom, Nick. 2014. *Superintelligenz. Szenarien einer kommenden Revolution.* Berlin: Suhrkamp.

Cantril, Hadley. 1940. *The Invasion from Mars: A Study in the Psychology of Panic.* Princeton, NJ: Princeton University Press.

Denning, Kythrya. 2013. Impossible Predictions of the Unprecedented: Analogy, History and the Work of Prognostication. In *Astrobiology, History, and Society. Life Beyond Earth and the Impact of Discovery,* Hrsg. Douglas A. Vakoch, 301–312. Heidelberg: Springer.

Dick, Steven J. 2013. The Societal Impact of Extraterrestrial Life: The Relevance of History and the Social Sciences. In *Astrobiology, History, and Society. Life Beyond Earth and the Impact of Discovery,* Ed. Douglas A. Vakoch, 227–257. Heidelberg: Springer.

Dick, Steven J. 2014. Analogy and the Societal Implications of Astrobiology. *Astropolitics. The International Journal of Space Politics & Policy* 12: 210–230.

Engelbrecht, Martin. 2008. Von Aliens erzählen. In *Von Menschen und Außerirdischen. Transterrestrische Begegnungen im Spiegel der Kulturwissenschaft,* Hrsg. Michael Schetsche und Martin Engelbrecht, 13–29. Bielefeld: transcript.

Fink, Alexander, und Andreas Siebe. 2006. *Handbuch Zukunftsmanagement. Werkzeuge der strategischen Planung und Früherkennung.* Frankfurt am Main: Campus.

Halbwachs, Maurice. 1967. *Das kollektive Gedächtnis.* Stuttgart: Enke.

Harrison, Albert A. 1997. *After Contact. The Human Response to Extraterrestial Life.* New York, London: Plenum Trade.

Harrison, Albert A., und Joel T. Johnson. 2002. Leben mit Außerirdischen. In *S.E.T.I. Die Suche nach dem Außerirdischen,* Hrsg. Tobias Daniel Wabbel, 95–116. München: Beust.

Hurst, Matthias. 2008. Dialektik der Aliens. Darstellungen und Interpretationen von Außerirdischen in Film und Fernsehen. In *Von Menschen und Außerirdischen. Transterrestrische Begegnungen im Spiegel der Kulturwissenschaft,* Hrsg. Michael Schetsche und Martin Engelbrecht, 31–53. Bielefeld: transcript.

Kuiper, T. B. H., und M. Morris. 1977. Searching for Extraterrestrial Civilizations. *Science* 196: 616–621.

Lowric, Ian. 2013. Cultural Resources and Cognitive Frames: Keys to an Anthropological Approach to Prediction. In *Astrobiology, History, and Society. Life Beyond Earth and the Impact of Discovery,* Ed. Douglas A. Vakoch, 259–269. Heidelberg: Springer.

Michaud, Michael A. G. 1999. A Unique Moment in Human History. In *Are we Alone in the Cosmos? The Search for Alien Contact in the New Millennium,* Ed. Byron Preiss und Ben Bova, 265–284. New York: ibooks.

Michaud, Michael A. G. 2007a. *Contact with Alien Civilizations. Our Hopes and Fears about Encountering Extraterrestrials.* New York: Springer.

Michaud, Michael A. G. 2007b. Ten Decisions that Could Shake the World. http://avsport. org/IAA/decision.pdf.

Pooley, Jefferson D. 2013. Checking Up on The Invasion from Mars: Hadley Cantril, Paul Felix Lazarsfeld, and the Making of a Misremembered Classic. *International Journal of Communication* 7: 1920–1948.

Schetsche, Michael. 2003. Soziale Folgen der Entdeckung einer außerirdischen Zivilisation (dreiteilig). *Nachrichten der Olbers-Gesellschaft,* Teil 1, Heft 200 (Januar 2003): 33–37; Teil 2, Heft 202 (Juli 2003): 26–30; Teil 3, Heft 203: 7–11.

Schetsche, Michael. 2004. Der maximal Fremde – eine Hinführung. In *Der maximal Fremde. Begegnungen mit dem Nichtmenschlichen und die Grenzen des Verstehens,* Hrsg. Michael Schetsche, 13–21. Würzburg: Ergon.

Schetsche Michael. 2005a: Zur Prognostizierbarkeit der Folgen außergewöhnlicher Ereignisse. In: *Gegenwärtige Zukünfte. Interpretative Beiträge zur sozialwissenschaftlichen Diagnose und Prognose,* Hrsg. Ronald Hitzler und Michaela Pfadenhauer, 55–71. Wiesbaden: Springer VS.

Schetsche, Michael. 2005b. Rücksturz zur Erde? Zur Legitimierung und Legitimität der bemannten Raumfahrt. In *Rückkehr ins All* (Ausstellungskatalog, Kunsthalle Hamburg), 24–27. Ostfildern: Hatje Cantz.

Schetsche, Michael. 2008. Der maximal Fremde – eine Hinführung. In *Der maximal Fremde. Begegnungen mit dem Nichtmenschlichen und die Grenzen des Verstehens,* Hrsg. Michael Schetsche, 13–21. Würzburg: Ergon.

Schetsche, Michael, René Gründer, Gerhard Mayer, und Ina Schmied-Knittel. 2009. Der maximal Fremde. Überlegungen zu einer transhumanen Handlungstheorie. *Berliner Journal für Soziologie* 19 (3): 469–491.

Steinmüller, Angela, und Heinz Steinmüller. 2004. *Wild Cards. Wenn das Unwahrscheinliche eintritt.* Hamburg: Murmann.

Uerz, Gereon. 2006. *ÜberMorgen. Zukunftsvorstellungen als Elemente der gesellschaftlichen Konstruktion der Wirklichkeit.* Paderborn, München: Fink.

Ulbrich Zürni, Susanne. 2004. *Möglichkeiten und Grenzen der Szenarioanalyse.* Stuttgart, Berlin: WiKu.

Consequences of First Contact—A Scenario Analysis

Following the preliminary methodological considerations, we will present four exemplary individual scenarios in the following—in accordance with the outline of the basic scenarios presented[1]: three of them describe and analyse the social consequences of distant remote contact (*signal scenario*), near remote contact (*artifact scenario*), and near close contact (*encounter scenario*). Our explanations will follow an identical representational logic in each of these three cases: After a description of the general parameters of the scenario, we examine the most likely effects on some important social fields (public, science, politics, economy, religion). All scenarios, as we have already indicated, are likely to have worldwide effects in a globalized culture. To describe them systematically here is simply impossible for more than one reason. In each sub-chapter, therefore, we will concentrate on exemplary forecasts that take a look at the consequences we have assumed in one or another part of the world—not always, but often Europe, because we simply know more about the 'conditions' there. In this sense, the following forecasts are a little 'Eurocentric' here and there for informational reasons. Finally, for each scenario, the abstract explanations are supplemented by an *exemplary narrative* in which we attempt to condense the respective initial parameters

---

[1] Baum et al. (2011) present a completely different scenario analysis. They assume a wide variety of interests and basic ethical attitudes of conceivable extraterrestrial civilizations and attempt to forecast what consequences this would have for human society in the event of first contact. Accordingly, three basic scenarios are distinguished: "Beneficial", "Neutral" and "Harmfull" (all from the human point of view). To us, such an analysis seems problematic, at least in detail, because ultimately unfounded anthropocentric presuppositions have to be made: The known interests and basic attitudes of human societies are inferred to those of extraterrestrial intelligences. Nevertheless, the analysis of the group of authors is *grosso modo* insightful, because it demonstrates the tremendous broad spectrum of possible processes in an interstellar cultural contact. Some general considerations on conceivable scenarios of first contact can also be found in Michaud (1999, pp. 208–218).

© The Author(s), under exclusive license to Springer Fachmedien Wiesbaden GmbH, part of Springer Nature 2023
A. Anton and M. Schetsche, *Meeting the Alien*,
https://doi.org/10.1007/978-3-658-41317-0_8

and the most important of the dimensions we have addressed into short science fiction-like narratives. They are exemplary in the sense that, in order to be able to narrate them in this form, we make a number of additional determinations, regarding the initial situation. In addition, we find various narrative elements that merely serve to condense the description, but have no systematic meaning (such as dates or names of companies). It should be clear that other concrete initial situations and deviating processes are conceivable in each of the scenarios examined. We assume, however, that the question of how the respective contact came about *in detail* and how it developed in the following makes only a partial, but not a fundamental, difference in the analysis of the general cultural effects of the initial contact of the respective species. Each of these subchapters concludes with a summary assessment of the respective scenario. A further chapter summarizes our predictions, regarding the different scenarios, in a clear form.

All this is supplemented by a concluding excursus dealing with the possibility of the 'failure' of first contact. This additional *fourth* scenario is dedicated to the case that the knowledge about the existence of extraterrestrial intelligence, which is accessible in principle, is scientifically and publicly rejected and remains part of the disputed field of *cultural heterodoxies* for a long period of time. We have chosen the *necrology scenario* for this purpose, which, like the artifact scenario, belongs to the field of what we call distant near contacts.

8.1   The Signal Scenario

8.1.1   General Parameters

(1) This is the scenario that also underlies the SETI programs that have already been reported on (Chap. 4): Radio telescopes or other technical equipment catch signals from the vastness of space that are of artificial origin. From the technical parameters of the transmission origin coordinates and the approximate distance of the transmitter can be deduced as well as perhaps something about its technical possibilities (Harrison 1997, pp. 199–200; Shostak 1999, pp. 231–232; Harrison and Johnson 2002, p. 100; Hoerner 2003, p. 133). In addition, at least if the signal is prolonged or if it is repetitive, it is likely that something can be said about the source's proper motion. Is the source in orbit around a sun, or is it moving slower or faster in space toward or away from the Earth (we used the identification of a Doppler effect in the signal to specify its origin in the exemplary narrative below). This information is especially important because it can be used to decide whether the transmitter

is a planetary device or a space probe or spacecraft traveling in the vastness of space.

(2) If the signal is modulated in one way or another in such a way that it is likely to contain a message, that content must be distinguished from information that can more or less be inferred from the physical nature of the signal itself. What information this might be is still a matter of debate in SETI research (Schmitz 1997). Although various considerations have been made about the decoding of radio or laser signals (e.g. Freudenthal 1960; Fuchs 1973, pp. 47–93; McConnell 2001, pp. 181–346; Elliott 2009, 2010 and 2011), they are all based on human thought patterns and cannot be transferred to the aliens without presupposition (Shostak 1999, pp. 232–233; Finney 1990). In them, it is unquestionably assumed that aliens would encode a signal just as we do, and that we could decode their signals just as they could decode ours. However, this is unlikely even if we are not merely listening in on a message intended for others, but are receiving an intentional *contact signal* (Hoerner 1967, p. 14). It could therefore be that we will not be able to decode the content of the received signal for a long time or perhaps even permanently.[2] At least in this case Marshall McLuhan's dictum applies quite literally: "The Medium is the Message". This too must be taken into account when assessing the consequences of this event.

(3) It is possible that in this form of first contact we do not learn much more than the fact: We exist. The more correct formulation here is: We *existed*. The interception of a light-fast signal that has travelled 1,000 light years also means that this signal was emitted 1,000 years ago. On the one hand, this connection is important in relation to the idea of a *cosmic dialogue*. On the other hand, the size of this distance is probably also important in a double sense for the concrete reactions on Earth: The greater the space and time that the message had to cover, the more indifferent the psycho-social reactions on Earth are likely to be: "Distance is critical because it structures the nature of the contact [...] the closer the contacting civilization, the greater the impact" (Michaud 2007, p. 211).

---

[2] "Deciphering a message could take a long time. Indeed, it could take forever. As we've mentioned before, any signal we're likely to detect will come from a civilisation that is enormously more advanced that we. The aliens' message might be impossible to unravel" (Shostak 1998, p. 194).

8.1.2   Projected Impacts

Let's start right away with this question of distance: A greater distance of the transmitter from Earth (we estimate that a socio-psychological limit here might be 100 or 200 light years) would move the signal far out of the human relevance system in terms of time, but also in terms of space. A signal from, say, 1,000 (light) years away would primarily be tangential to Earth's scientific, philosophical, and religious subsystems, but would be rather irrelevant to people's lives and everyday consciousness. Initially, there would certainly be a great deal of *public interest* and also discussion in a wide variety of *mass and network media*.[3] However, in view of the small amount of information received (assumed here) and because of the impossibility of an immediate dialogue, the topic would quickly disappear from the public eye. Various scientific programmes would certainly be launched to extract the maximum amount of information from the signals received, and perhaps also those dealing with the question of whether and how to respond. Not least because of the monologue situation and the long periods of broadcasting, most people's interest in such 'contact' is likely to wane very quickly. Consequences of signal reception would therefore, we believe, be primarily for religious-philosophical systems and for the sciences (see also Shostak 1998, pp. 189–198, who takes a similar view).

It would be different if the signal came from the immediate *cosmic neighbourhood* of the earth. The distance threshold to be assumed here, from a socio-psychological point of view, would probably lie in the range corresponding to the lifetime horizon of the average members of our society, i.e. thirty or at most perhaps one hundred (light) years. Only within this time period and thus also this distance would alien civilizations be perceived cognitively as well as emotionally as 'reachable neighbors'—reachable both in terms of the possibility of civilizational dialogue by means of radio waves or other signals, and in terms of the theoretically conceivable possibility (on Earth today) of direct contact in the foreseeable future. The consequences of such a 'middle-distant contact' could consist—apart from the religious and philosophical change of world view, which is rather independent of the distance of the alien civilization—on the one hand

---

[3] Jones (2013) examines expected differences and similarities in reporting and discussion between traditional mass media and the new social media following the reception of an extraterrestrial signal. His conclusion: "We live in a media-saturated world. News of an extraterrestrial discovery would travel quickly and somewhat haphazardly through the mainstream and social media. Dubious information would certainly appear amid the flotsam of messages on the subject. It would be circulated extensively, but would not necessarily have much influence in the face of more credible reports" (Jones 2013, p. 327).

in a change of national and international strategies of research policy in order to realize a corresponding dialogue or even direct contact. On the other hand, *fears* about the consequences of contact in the terrestrial population are also likely to increase as the distance decreases.

Regardless of the distance of the signal source, the *political and economic effects* of this scenario should ultimately remain rather insignificant.[4] On a political level, the probability that an extraterrestrial power—possibly technically and therefore militarily far superior—will interfere in earthly affairs remains extraordinarily low. The dictum of SETI research, THEY could not reach here because of the great distance in space, remains in force and certainly exerts a calming effect in the field of politics.[5] For the sovereignty of the earthly nation states there is no danger[6]—therefore everything speaks for the fact that this remote contact will have no structural effects on the political structure on earth. On the other hand, the political developments on Earth, after the signal reception, seem to us to be very significant for the question regarding public attention the signal will receive from the world's public in the medium and long term. The more political, but also ecological and economic crises that shake the public in the following years, the faster the signal reception will be forgotten.

The only exceptions are those *scientific disciplines that* can contribute to the investigation and, in the long term, perhaps even to the decoding of the signal. In addition to astrophysics, these are likely to be computer science, linguistics, cognitive science—and possibly then also exosociology. In countries with the appropriate research resources, comparatively extensive research projects are likely to emerge that will work on the study of the signal for years or decades. The flow of appropriate research resources is likely to be more intense and last longer the more durable the signal is picked up and the more complex it is. In this context, it seems likely that in a number of countries, government funding is now being made available for *new SETI programs* whose mission is to search for more extraterrestrial signals. A real scientific breakthrough would only be recorded if it were really possible to decode the signal in question. However, if one follows the

---

[4] A different position is taken by Billingham et al. (1994, p. 87), for example, who assume that even an undecipherable extraterrestrial signal would have considerable consequences for the world view and also for the political situation on Earth.

[5] Typical of this are, for example, the remarks of Billingham et al. (1994, pp. 61–120), who deal exclusively with the possible reactions of the public and the political-administrative system to the reception of an extraterrestrial signal—other possible contact scenarios are tacitly omitted.

[6] Cf. Wendt and Duvall (2008), who systematically deal with the opposite case—we will come back to this in the context of the encounter scenario.

basic objections of communication science as formulated decades ago by Russian researchers (see exemplarily Sukhotin 1971; Panovkin 1976; Sheridan 2009, pp. 67–103 and Hickman and Boatright 2017), the decoding of an extraterrestrial radio message should be unlikely, perhaps even impossible in principle.

But even if the alien message could be decoded in principle after perhaps decades of effort, it seems more than questionable what understanding of the aliens, their culture, science and technology would actually result. As Sheridan (2009, pp. 141–166) comprehensibly points out, there now exists a wealth of quite plausible assumptions from evolutionary biology and cognitive research about the emergence of an entirely human-unlike intelligence whose messages can hardly be correctly interpreted by us, even if we understand them superficially in a linguistic sense. However, this also means that the effects of a sudden scientific and technological thrust due to the decoding of the alien message, which are often hoped for in SETI research, will fail to materialize. Accordingly, we assume that the technological and resulting economic effects of remote contact will remain rather small.

Possibly, the primary consequences of a first contact would therefore be located more in the *cultural and religious field*[7] than in the field of technology, economy and politics. In the medium and long term, the reception of the signal could lead to a rethinking on Earth, with regard to the position of humans in the cosmos. The certain knowledge that, as an intelligent species, we are not alone in the universe could provide lasting motivations for the development of, for example, *transhuman ethics*, in which thought is given to the treatment and rights of non-human entities. Long before closer contact with an extraterrestrial species can be realized, this could already benefit entirely terrestrial entities: various animal species (especially primates and dolphins) as well as the AI-controlled robots or androids that increasingly dominate everyday life. Similar long-term effects also seem conceivable in the field of religion and spirituality (although we believe that developments here are extremely difficult to predict): Transhuman-oriented religious communities could emerge, according to whose conviction not only humans but more broadly all intelligent beings could be subject to spiritual development or even divine grace. We consider the development of such comprehensive communities to be more promising in the long run than the emergence of various

---

[7] The consequences of a received extraterrestrial radio message for the Christian faith and Catholic theology are discussed by Ascheri and Musso (2002). Peters (2013) also asks about the religious consequences of a signal reception, and immediately argues for the formation of a new discipline, which he calls "astrotheology"; at the center of his considerations is the question of whether the certain knowledge of the existence of extraterrestrials would trigger spiritual or religious crises on earth.

small cults which interpret the extraterrestrial signal as in some sense divine or transcendent. Such groups will probably exist in the first years after the signal reception, but in our estimation, they will hardly survive the waning of the general cultural interest in the subject.

8.1.3 Exemplary Narration

In May of the year 2025, a not very strong but tightly focused radio signal from interstellar space is received in China: Within 11 days, a short signal (lasting just under 5 seconds each) with extremely complex modulation reaches terrestrial radio telescopes three times. Unfortunately, the source of the signal cannot be determined exactly; based on various physical parameters, it is estimated to be between 500 and 2,500 light-years from Earth. No Sun-like star can be identified in the direction of the signal at this distance. The three signals differ noticeably in wavelength—probably a Doppler effect, which would mean that the signal source is moving away from the Earth with *increasing* speed (i.e. accelerated by some force). This would then suggest, though this is a controversial inference, that the signals were emitted many centuries ago by a space probe rather than a planetary station. This question will occupy the international scientific community for a long time—but only them.

The first signal is picked up only by the large Chinese radio telescope *Tianyan*—due to a breaking news via international professional societies, other telescopes on Earth are aligned and calibrated accordingly, so that the two subsequent signals can be detected by 4 and 7 other terrestrial telescopes, respectively. Even before the reception of the third signal, the *SETI Post Detection Protocol* is activated; according to the provisions of the "Declaration of Principles Concerning Activities Following the Detection of Extraterrestrial Intelligence" (of the International Academy of Astronautics) from 1989, various international scientific organizations are informed. After the reception of the third signal, while the coordination process between the various research institutes and international professional societies is still underway, political decision-makers of the countries with large radio telescopes as well as the UN Secretary-General and the World Security Council will be informed.

Contrary to the protocol and internal agreements between the involved observatories, two research groups from different countries go public in hastily organized press conferences shortly after the reception of the third signal within a few hours of each other. They justify this step with the fact that, on the one hand, there had been increasingly intense speculation about the reception of strange signals from space in the social media area and on some Ultranet platforms for days, and that, on the other hand, it would be irresponsible not to inform the world public immediately about this truly groundbreaking development. The press conferences lead to a fierce, but in the end quite short media echo. In Germany, tabloid newspapers have their say with headlines like "E.T. phones home"—other German leading media take the opportunity to once again make fun of the waste of money of the SETI projects and the perverse motives of the research teams involved. ("Alien hunters on the sidelines" is the title story of a weekly magazine dominating public opinion). At the Federal Press Conference on the topic of "Future Perspectives of Research Policy", the Federal Minister for Education, Research and Technology, in response to a journalist's question to this effect, explains that the Federal Government continues to refuse to waste resources on "purely speculative projects that have no discernible benefit whatsoever for the people in our country".

However, due to the increasing political tensions in Southeast Asia, which lead to a devastating second Korean War in the late summer of 2025, public interest in the radio signal fades as quickly as it had arisen. The research teams involved -it becomes clear at an international conference (which, however, receives little public attention, not least because of the devastating summer drought of 2026 in much of Africa) a good year after the first reception -are largely in agreement that it was an artificial signal from an extraterrestrial intelligence. Since the signal was received only three times, was very short and extremely complex modulated, the experts assume that a decoding would take many years, perhaps even never be possible. This is probably one of the reasons why the scientific debates and also the—increasingly rare—public debates will shift more and more in the coming months and years away from the question of which message was possibly transmitted by whom and to whom, towards the question of when is generally the right time to inform the public about the reception of a possible extraterrestrial signal. In this context, the "Post Detection Protocol", which dates back to the 1980s, is also strongly criticized because of its cumbersomeness and the resulting time delays. Some of the research groups involved in the signal reception at that time (among them also the Chinese researchers) clearly indicate that they will go public much faster 'next time' of their own accord—all international agreements on a controlled information policy in case of a signal contact seem to be invalid.[8]

---

[8] At the end of 2022, The SETI Post-Detection Hub was founded at the University of St Andrews (UK) with the goal of developing protocols for such scenarios. The homepage

8.1.4 Summary Assessment of the Scenario

In general, our assessment is that the further away the sender of the received signal is and the longer the decryption attempts of the sent content last, the more the public interest in this kind of first contact will wane and the smaller the medium and long-term cultural effects will be. The only exceptions to this are likely to be those scientific disciplines directly concerned with decoding possible content and searching for further signals. A first contact of this kind is only likely to have technological and economic effects if it should actually succeed (against all probabilities) in decoding the contents of the signal in a linguistic sense. Whether this includes the interpretive understanding of the signal in a psychological or sociological sense is quite another question. It is conceivable that the event could trigger cultural developments in the philosophical and religious-spiritual spheres.[9] The probability of this is difficult to assess. Not only this question depends, in our estimation, on the political, economic and ecological developments on our planet in the following years and decades, which are independent of the event. The more dramatic these developments will be, the more completely terrestrial crises will afflict mankind, the less interest there will be in the possible decoding of the signal and probably in the further search for extraterrestrials. The greatest impact outside of the sciences involved in this research process is likely to come from this remote contact in the realm of artistic and, in the broadest sense, cultural representation: Numerous new novels and films, television series and interactive media are likely to emerge whose plots take their starting point in the reception of an extraterrestrial signal. However, the very real everyday life of humans will not change.

---

states: "The time is thus right for consideration of humanity's response—and responsibility—following the detection of both life and intelligence in the Cosmos. We should plan now for this eventuality by setting out impact assessments, protocols, procedures and treaties designed to allow humanity to respond responsibly." (https://seti.wp.st-andrews.ac.uk/).

[9] In contrast, Baxter and Elliott (2012, p. 34) are of the opinion that a received extraterrestrial message can be so serious or risky in terms of its content that it could become necessary to destroy the message itself and all associated data. We have doubts about this thesis already because in our opinion it should be as good as impossible to decipher the contents of an extraterrestrial message.

8.2   The Artifact Scenario

8.2.1   General Parameters

(1) The artifact scenario (Harrison and Johnson 2002, p. 113; Michaud 2007, pp. 135–140) describes the situation in which humans, while exploring outer space, at some point encounter the material legacies of an alien civilization in the nearer or more distant vicinity of Earth.[10] For such a find to have any cultural consequences, two conditions must be met: The object found on Earth or in near-Earth space (or, in the future, somewhere in our solar system) must—according to the scientifically and/or socially *dominant interpretation*—firstly not come from Earth and secondly certainly be of artificial origin. The two prerequisites are realized within the scope of different degrees of interpretation: While the extraterrestrial origin of an artifact— under the widely accepted assumption that the Earth is today experiencing its first space age—can already be reliably deduced from a site of discovery *beyond the Earth*,[11] the problem of the naturalness or artificiality of a corresponding object could arise all the more persistently the further the technical capabilities of the original culture reach beyond those of the terrestrial one (or at least deviate from it). Conceivable are objects of such strangeness that not only every method of technical investigation known today fails,[12] but that even the classification according to the guiding difference 'artificial' versus 'natural' remains doubtful.

(2) The latter is especially true if the object found does not have any symbols of foreign origin that are recognizable to humans. But even if 'inscriptions' are found, scepticism about their decipherment is in order: even in the case of human cultures, unknown writing, as long as there are no reference sources,

---

[10] It is known in its fictional form, for example, from the film *2001: A Space Odyssey* (UK/ USA 1968, directed by Stanley Kubrick). In this tale, humans exploring the moon come across the legacy of an alien civilization, an artifact in the form of a black monolith that was apparently left there several million years ago for the purpose of making future contact. After an unintentional activation by the humans, the monolith begins to emit an automatic message into the depths of space (Hurst 2004, p. 104).

[11] In the case of finds made on Earth, it would be much more difficult to establish an extraterrestrial origin beyond doubt. Here we would be with the theses of the so-called paleo-SETI-research, which we will deal with in more detail in Sect. 11.1; we therefore exclude this special case at this point.

[12] A good example from science fiction here is the alien *Sphere* in the feature film of the same name (USA 1998, directed by Barry Levinson).

presents science with insurmountable problems of interpretation. For corresponding legacies of past terrestrial cultures, such as the 'Phaistos Disc', there are a multitude of competing interpretations, but to date there is no scientifically accepted decipherment of the inscription (Duhoux 2000). This problem presented itself to an incomparably greater extent with extraterrestrial artifacts. Even a rich symbolic decoration of a found artifact would probably not provide any information about the thought structures or even the motives of the extraterrestrial 'authors'.[13] Since the place where the artefact was found and the situation in which it was found probably only allow for a few certain conclusions (because we know nothing about the motive structure of the aliens, we cannot draw conclusions about possible motives of the others on the basis of our human considerations—which seems to us to be well considered, may in fact have been thrown away haphazardly and vice versa), the consequences on Earth will primarily result from the fact of the find itself, i.e. from the interpretation of the object as being produced by extraterrestrial intelligences, instead of from a possibly never to be deduced meaning and functionality for its original creators.

(3) How massive the cultural impact of such a discovery would be depends, in our estimation, primarily on two factors: (a) A possible age determination would move found objects into or, on the contrary, out of the human time horizon. An estimated or calculated age[14] of one hundred years would have a completely different meaning here than one of ten million years. In the former case we would be confronted with immediate 'temporal neighbours' who, if the find took place in the immediate cosmic neighbourhood of the earth, would probably indicate the existence of a civilisation on earth. In the latter case, however, all such considerations would be self-explanatory (Michaud 2007, p. 212). (b) If one of the objects found could be interpreted as the material basis of some kind of technology, this would immediately lead to speculation about the nature of that function(s) and certainly about

---

[13] It remains to be asked, however, what difference it makes for information retrieval whether an artifact was intentionally left behind for the purpose of communication or whether it represents a more accidental legacy. Freitas and Valdes (1985), for example, distinguish three basic cases: (I) artifacts are intended for contact, (II) artifacts avoid discovery, and (III) the question of discovery by a third party is irrelevant to the artifact's abandonment. Assuming the author's assumption that artifacts in the second case are in fact undetectable due to their advanced technology, only Case I and Case III would need to be analytically distinguished here.

[14] The scientific question of the possibility or impossibility of determining the age of the object in question is left aside here.

the current functionality of the object in question. The question would then be not only what the artefact can do, but in particular what consequences this might have for its immediate or wider environment. This entails a whole series of serious practical questions in the aftermath of the find: should the object be left as untouched as possible, or should it be systematically studied scientifically? Can and should it be transported to another location, possibly even brought down to earth from space? Should it—if technically possible— be manipulated in some way or even disassembled (if this is at all possible with our resources? And other things.

8.2.2   Projected Impacts

If we assume the discovery of one or more extraterrestrial artifacts on an asteroid not too far from Earth (we also use this case in the exemplary narration in the following subchapter), the consequences of the discovery, if it becomes public, on the general world view of the population are considerable, at least in scientistic societies. In this scenario, a stronger cultural 'impact,' compared to the already discussed remote contact, results from the fact that its 'it-gives-us-message' is not only overlaid by a 'we-were-here-message', but also dominated scientifically and psychosocially (Michaud 2007, p. 211). With such a discovery at this moment in time, all the theses (which have dominated the field of traditional SETI research to date) about the unbridgeability of interstellar distances through space travel would be exposed at a stroke as an anthropocentric prejudice. It would be proven that other civilizations are very well able and willing to bridge the corresponding distances at least with automatic space probes.

The discovery of an extraterrestrial artifact in our solar system would there-fore, in our estimation, arouse great interest not only in the scientific world, but also in the general public, at least in industrial and post-industrial society. An immediate consequence would probably be that all space-faring nations and many space companies would make great efforts to find further extraterrestrial artifacts in the solar system. The SETA projects (we reported on them in Chap. 5), which are still mostly ridiculed both publicly and scientifically today, would experience a major upswing; the technical and financial resources expected to be deployed there would probably also revolutionize solar system exploration in general. We therefore assume that a discovery of this kind in the Solar System, regardless of all the specific details, would provide a very strong impetus for unmanned space-flight, and perhaps also for manned spaceflight. Such a discovery could well be the beginning of a very intensive phase of scientific-technical exploration and in a certain way also (commercial and nation-state) 'appropriation' of large parts of the solar system.

In the *scientific field*, moreover, the find would certainly trigger a whole series of research projects aimed at 'extracting' maximum information from the object or, if it is a larger complex artifact, from the objects. This would certainly involve various disciplines: Astrophysics would try to find out something about the origin of the object. Chemistry, physics and material science would take care of the composition. If there is even a hint of functionality, it would be the task of various engineering disciplines to find out more. And if symbols can also be found on the object, this would be a task for linguistics. Comparative anthropology as well as cultural and social sciences could also make their contribution, when it comes to the questions with which technical means and also with what kind of extremities such an object was built; what was or is its purpose and what we can deduced from it, about the culture of its creators. The more the investigations go in this direction, the more speculative and lengthy they would become. How great the overall research effort would be depends, of course, also on the size of the object and the place where it was found: Must (or even: should) it be examined on site, or can the whole object or at least parts of it be brought to earth?

This leads us to the last question and into the realm of *politics*. Here, a central point is likely to be the first issue: Who has the power of disposal over the object—legally, but especially also quite factually? We assume that after the find (if it cannot be recovered immediately and brought down to earth before any international discussion, and is then factually owned by a nation or corporation) there will be fierce political disputes over the 'rights' to the object or group of objects. To date, the provisions of international space law are not specific enough to set even the most general standards for the case of such an artifact discovery (Schrogl 2008; Baxter and Elliott 2012, p. 34; Gertz 2017, pp. 3–4). For extraterrestrial artifacts found outside of Earth, there are so far no international legal regulations. But even if these benchmarks and more concrete regulations were in place at the time of the artifact's discovery, it seems to us highly questionable whether multinational corporations or powerful nation-states would feel compelled to abide by such provisions of international law. The immense number of interventions and acts of war all over the world in recent decades, which undoubtedly violate international law, shows abundantly clearly that international law is worth virtually nothing in practice, when militarily powerful nation states wish to assert their own interests. We therefore anticipate that the discovery of an artifact believed by various parties to have technological potential value could lead to high-risk international conflicts: conflicts that could well include military options. If a found object cannot be hastily recovered and brought to Earth, it seems to us that not only is a race among state and private space actors to gain access to the project possible-we would not rule out the possibility of a military confrontation in space if actors hope to secure sole access to the artifact(s) by threatening or even using armed force. But even if a corresponding object can be quickly brought to Earth, it remains a further danger as a potential trigger for war—for example, if investigations reveal that the object contains usable technology in a quantity or quality that would enable the corresponding owner to make a virtually unassailable technological advance. Here, from secret operations to open warfare, all strategies are conceivable to gain possession of the corresponding 'alien technology'.

The *public* would certainly accomapny this with varying amounts of attention. How the necessary change of world view is dealt with here and what effects this has on public opinion certainly depends strongly on the parameters already mentioned: Public interest is likely to be greater the younger the object and the more successful it is in attributing certain functionalities to it. However, these two factors also determine the question of what 'emotional colouring' is given to public interest. A relatively new object suggests the question of whether its creators might return in the near future, which could well fuel a whole range of sometimes more, sometimes less well-founded fears among the public. If the object has a recognizable, still intact technical functionality, the question arises what this could do and how it can be influenced. And, of course, there is also the question of what consequences this or that function could have for its surroundings. The result would most likely be heated (scientific, but also political and public) discussions about the possibility and necessity of attempting to manipulate the corresponding 'mechanism' and thus influence the alien functionality. The matter would be even more complicated if the object had a recognizably running action program that would react to changes in its environment (such as human activities) with a perceptible change of state.

Yet a scenario like that in *2001: A Space Odyssey*, in which the artefact is activated by human manipulation and subsequently sends a message into the vastness of space, is only superficially the 'worst case'. It does seem ominous because humanity is stripped of the active decision to make contact, but leaves no doubt that something potentially consequential has happened. Even more problematic in terms of mass psychology seems the case in which it remains unclear whether, after an activation—intentional or unintentional—by the human finders, a message has been sent to the producer of the object on channels undetectable to us. It is precisely this unknown that, as a permanent potential threat, could become highly effective culturally in more than one respect: in the field of visual art and fictional representations, with regard to the change in people's basic collective psychological constitution, or also through the medium- and long-term change in the political agenda. The possible effects here are extraordinarily diverse, but difficult to assess in detail, because analogous earthly cases (of asymmetrical cultural contact mediated exclusively through alien artifacts) may have occurred several times historically, but to our knowledge have not been systematically studied in terms of cultural history.

The prognosis of *religious-spiritual* and economic consequences of the event seems to us to be similarly problematic. The former consequences are likely to be similar to those of signal-mediated first contact in the scenario described earlier. It seems likely to us, however, that in this case the number of emerging space-related cults and their appearance would be more vehement, since the option is clearly stronger in the 'room' that technically superior and thus (in naive interpretation) more 'god-like' visitors will once again 'return' to our solar system or even to Earth and carry with them transcendent knowledge or corresponding enlightenment, in which they let us (hoped for by the believers) participate.[15] Here, too, the emergence of a multitude of sectarian groups is to be expected, with the aim of establishing contact with the creators of the artefact—or even claiming to have already done so in this or that way (for which pseudo-scientifically dressed-up as well as entirely esoteric 'communication channels' such as astral travel to the stars or telepathic collective signals come into question). The public will take note of all this, partly with interest, partly with amusement. For many followers of such cults and groups however, a completely new 'sense of life' might come up for a shorter or longer time. In this respect, too, an artefact discovery could therefore be extremely meaningful in the medium and long term.

Whether it is also meaningful and thus also profitable in the field of *economics is* likely to depend, on the one hand, on whether technologically usable information can be extracted from the artefact, as it were—and of course on whether companies or at least national groups of companies succeed in monopolizing access to this information.[16] All this is difficult to assess, not least because

---

[15] In this context, public attention and enthusiasm for the theses of pre-astronautics should also increase very strongly: If aliens left artifacts somewhere in the solar system, why couldn't they have left their mark on Earth? Although we are not ourselves adherents of pre-astronautical thinking, we must concede that an artifact discovery within the solar system would also force the sciences to take a serious look at this line of thought, which remains heterodox to this day.

[16] Billingham et al. (1994, pp. 84–87) already warn of the conflicts that could be triggered by attempts to monopolize the information obtained about extraterrestrials on a nation-state basis—although they refer exclusively to a signal scenario.

it is impossible to say in advance whether any information that can be used in this way (i.e. technologically and economically) can be extracted from the object. It could well be that the objects are so alien or that the technology used is so advanced that it is impossible for us humans to extract even the smallest usable technological information from them in the foreseeable future. We may lack any scientific basis to even begin to understand what the objects we find are doing and how they do it.[17] It is therefore much easier to predict that the artefact discovery will boost at least one terrestrial industry beyond measure: the companies specialising in the exploration and exploitation of space. And those companies or consortia that have already participated in SETA projects and have the appropriate search and perhaps even recovery technologies will automatically play a pioneering role. It is therefore already decided before such an event who will have the economic edge after its occurrence.

8.2.3   Exemplary Narration

In May 2025, the commercial space probe "Vanguard 2" of the "European Asteroid Mining Corporation" unexpectedly discovers some artificial structures of several meters in size while mapping the near-Earth asteroid "Horus 2007b", which is almost 120 meters in diameter. The structures could be externally visible parts of propulsion segments, which would indicate that this asteroid was deliberately maneuvered into this near-Earth orbit—possibly to collect observational data every four years as it approaches Earth. The consortium, which is registered in Luxembourg, wants to keep this potentially groundbreaking discovery secret for now. However, one of the consortium's major shareholders, a multinational Internet company, does not feel bound by this agreement and publishes images of the enigmatic structures on several of its Ultranet platforms, bringing the space probe mission considerable attention within a very short time. This ensures that the discovery receives a great deal of public attention right from the start, which even grows into a veritable media hype for a short time. Especially in the interactive media of the Ultranet, not only reports and discussions are overflowing, but also artistic adaptations and reworkings. More and more alleged images of the artificial-looking structures are circulating, so that the original photos and short film sequences of the "Vanguard 2" space probe can

---

[17] To this day, within science fiction, the novel *Roadside Picnic* by the Strugazki brothers (1975) is one of the most convincing examples of how perplexed and even helpless the confrontation with advanced alien technology can leave humans.

no longer be distinguished from computer-generated imitations. Virtually, whole groups of unexplored asteroids are 'paved' with artificial structures (and several very successful virtual reality games are created, in which the exploration of a 'secret alien base' on an asteroid is at stake). The abundance of computer-generated imagery complicates factual public debate of the discovery, but also makes scientific assessment increasingly difficult when more and more creative re-creations of photographs and films of supposed 'alien artifacts' also appear in scientific forums. Not least this development makes it clear to economic and political decision-makers that certainty about the nature of the observed structures can ultimately only be obtained by a direct exploratory mission to that asteroid.

The "European Asteroid Mining Corporation", whose exploration probe had discovered the artificial structures, claims the right to the first closer examination of the asteroid –a right, however, that is neither covered by international treaties nor recognized by other companies or even nation states. Since, according to the 2023 Locarno Treaty, all materials recovered outside the Earth belong to the company (state or private) that was technically capable of recovering them, the crucial question is who can reach the asteroid first in a direct exploration mission.

As a result, commercial spaceflight companies and national and multinational spaceflight organisations are working frantically. Due to the eccentric orbit of "Horus 2007b", the asteroid can only be reached with reasonable effort every four years. Various space agencies and business groups are trying to set up smaller or even larger missions for the next approach in 2029. NASA promises the US government to bring forward the first manned asteroid mission with a new Orion capsule—actually planned for 2032—and to redirect it to "Horus 2007b". A few months later, however, NASA had to concede that a mission of this kind would not be possible until 2033 due to technical difficulties. Two years later this date has to be postponed again: A manned flight to the asteroid is now envisaged for 2037. This means that NASA is out of the running. The same applies to the Europeans. Five months after the discovery of the structures on the asteroid, a special meeting of the science ministers of the member states is being held in Bonn at the request of the ESA administration. Despite intensive three-day discussions, no agreement can be reached on the financing of a prompt automatic mission to the celestial body (a manned mission was out of the question from the outset for technological and, in particular, financial reasons). In the face of considerable resistance from a number of Eastern European countries, which in principle regard exploration outside the Earth's orbit as a waste of taxpayers' money, a 'coalition of the willing' finally agrees to send a small robotic probe to the asteroid in 2037. In this way, ESA shares NASA's fate.

Inspired by the discovery on that asteroid, the "International Space Cooperation" (ISC) is formed at the end of 2025 under the leadership of some multinational space and technology corporations, whose members have considerable technical expertise at least in unmanned spaceflight. To the amazement of most space experts, after the ultimately failed Bonn conference, France, Italy and Luxembourg withdraw from the ESA and become the first nation states to join the new commercial space organization (several other European and non-European states follow suit in the coming years). Within two years, the ISC develops a concept that is as sophisticated as it is ambitious, based on existing launch systems and space capsules from some of its member companies, to take six astronauts to the asteroid for several months of exploration as early as 2029. The project falls through at the last minute when a newly developed, high-powered launch system designed to carry the large crew habitat into orbit for the long-distance mission accidentally fails during a launch attempt from French Guiana and crashes into the sea. A next launch attempt could be made in spring 2033 at the earliest.

Instead, to the amazement of the world public and some experts, a manned spacecraft of the recently formed "Asian Space Alliance" (ASA) reaches the asteroid "Horus 2007b" in autumn 2029, where the four-member multinational crew immediately begins with intensive explorations of the mysterious structures. In the course of this, large quantities of apparently extraterrestrial technology can be recovered. The ASA was founded shortly after the discoveries of the "Vanguard" space probe by the People's Republic of China, India and some smaller Southeast Asian states. After months of secret negotiations, Japan and South Korea also join the ASA in mid-2026. A few weeks after the signing of the accession treaty, Chinese troops invade North Korea and overthrow the regime there. The nuclear weapons found there are destroyed under international control; a transitional government is set up whose central task is to prepare for reunification with South Korea. Whether the Chinese military intervention was part of a secret supplementary agreement to the Contribution Treaty between Japan, South Korea and the ASA, the world public will never know. What quickly becomes obvious, however, is that the discoveries made on "Horus 2007b" give the member states of the ASA a clear technological advantage over all other economies in the world. Four years after the first mission, the ASA establishes the first permanently manned outpost outside the Earth's orbit on the asteroid, whose primary task is to explore the obviously complex—but now inactive—technical facilities that an extraterrestrial intelligence must have left on the asteroid thousands of years ago. In the following years, the growing "Horus" station becomes the starting point for the search for further extraterrestrial artifacts in the solar system and for the exploration of interplanetary space.

8.2.4   Summary Assessment of the Scenario

The medium- and long-term consequences of the discovery of an extraterrestrial artifact in the various social subsystems depend, in our estimation, on various factors: Public interest and mass psychological consequences are primarily influenced by the age of the object and its possible functionality—the younger and the more functional the object is, the more interest, but also collective concern, the find will trigger. On the economic level, on the other hand, the decisive question is primarily whether technologically usable information can be obtained through the investigation. If this is not the case, only space-related companies are likely to benefit from public attention to the discovery and, in particular, political interest in further corresponding artifacts. Politically, in turn, the discovery is all the more explosive the less clear the factual (rather than legal) ownership is and the greater the technological gain the artifact(s) promises. The fact of a previous visit by extraterrestrial intelligence is likely to influence the political agenda at most in the medium and long term—unless the object can be shown to have reached our solar system no more than a century or two ago, or to have functionality that can be interpreted as direct contact with its creators. Philosophically and religiously, on the other hand, all of the above parameters seem to be of rather secondary importance. Here, the existence of an extraterrestrial artifact in the solar system speaks entirely for itself. This is probably also true for scientific disciplines traditionally responsible for 'extraterrestrial questions'. Whether the find can trigger a greater scientific impact beyond that will likely depend on what information beyond that evidence of earlier extraterrestrial presence in the solar system can be gained through closer examination. With respect to most of the parameters mentioned, the space of possibilities created by their imaginable manifestations is simply too confusing to arrive at much more concrete predictions of the consequences of such an artifact discovery.

8.3   The Encounter Scenario

8.3.1   General Parameters

(1) First contact in the form of a direct encounter occurs when a non-terrestrial spacecraft appears in the terrestrial atmosphere or in near-earth space and it can be assumed, on the basis of its flight manoeuvres or other actions, that

it is controlled by an intelligence.[18] What is to be understood by 'near-Earth space' in this context is likely to have been subject to significant technical and cultural change over the last few decades. It seems reasonable to us to start from an understanding that focuses on the accessibility and influenceability of the corresponding part of space by human action. That part of space which can already be reached by human space probes or even space ships today or which could at least be reached in the near future then appears as 'near Earth'. If we combine this with more astrophysical boundaries, this is likely to include *large parts of the solar system* by the beginning of the twenty-first century, at least as far as the orbit of the outermost known planet orbiting our own Sun (i.e. roughly the orbit of Neptune).[19] We interpret technical exploration by a man-made space probe as a precondition of *cultural appropriation*, which would make the relevant part of space collectively-mentally a kind of Earth's front garden. Independent of this determination, the question of how far the alien object approaches Earth is likely to have additional implications for its perception and psychosocial consequences. Other boundaries to be considered analytically here are likely to be, in particular, the Moon's orbit (the Moon being the only celestial body outside the Earth that has received human visitation on several occasions), an Earth orbit at the typical satellite altitude, and then certainly the Earth's denser atmosphere and the Earth's soil itself—the invasive crossing of any of these boundaries is likely to make the mass psychological consequences on Earth more problematic.

(2) on the other hand, the question of whether it is a matter of contact with a primary (biological) life form, its artificial representatives, or the emissaries of a secondary machine civilization is likely to remain unresolved, at least initially, but possibly for a long time. This is due to the fact that it might

---

[18] All other technical questions, such as the propulsion of such a missile, are not only of a purely speculative nature, but moreover irrelevant for a sociological understanding of the trans-social situation unfolding in the first contact.

[19] Since this is primarily a demarcation based on collective human perception, it seems irrelevant that a few space probes have already left this area—with one exception (the space probe "New Horizons", which reached the dwarf planet Pluto in 2015) only long after the completion of their actual exploration missions anyway.

be difficult, if not impossible, to distinguish these three cases (or even imaginable combinations) before the establishment of an intensive exchange of information.

> There is a considerable overlap between the cases of direct contact and smart probes; possibly a sufficiently advanced post-biological ETI would *be* the probe. Thus it may be the best to treat an encounter with a smart probe as a case of direct contact. (Baxter and Elliott 2012, p. 35; emphasis in original; cf. Michaud 2007, pp. 128–130)

Since we know nothing about the extraterrestrials in advance, we cannot form any empirically even rudimentarily validated ideas about their external appearance. Even if a human-sized entity were to emerge from an alien spacecraft (this is one of the classic first-contact scenarios of science fiction) after landing on Earth, it remains undecidable in advance whether it is an evolutionary biological being, an artificial emissary that more or less resembles the form of its creators on the outside, or an AI-controlled entity from the collective of a machine intelligence (cf. Elliott 2014). This question arises even less when we are confronted with a spacecraft from which no smaller secondary units separate. Whether this missile as a whole is an AI-controlled probe, whether it conceals further artificial or biological entities in its interior (i.e. whether it represents a spaceship in our human sense), remains unresolved until the aliens provide us with concretely interpretable information about it. So, at first we can only react very roughly to externally observable phenomena: The spacecraft follows this or that trajectory, it comes this way or that close to Earth, possibly sends some kind of signals on wavelengths we can receive, eventually perhaps even lands on Earth (or sends secondary missiles into our atmosphere).

The most important factor for the terrestrial prognosis of the consequences of this event, however, is, in our opinion, precisely not such—and other imaginable—manoeuvres of the alien spacecraft (we had already pointed out in the previous chapter that it is hardly possible to draw conclusions about the motives and interests of a maximum alien from its external actions), but the *human interpretations* of the alleged 'actions' of the aliens, *guided by cultural patterns of interpretation* (cf. Harrison and Johnson 2002, p. 104). When predicting the consequences of 'close proximity contact', we must not forget that almost all people are already more or less familiar with corresponding scenarios from *science fiction*. For more than a hundred years, science fiction has provided countless media variants of the encounter scenario. First contact as a fictional event has therefore already been played out thousands of times in cultural thinking and has certainly also left its mark on collective thinking—in both a positive and negative sense. When analysing the cultural consequences of the event, it can therefore be assumed that almost all people have already been mentally confronted with a similar situation through their involvement with novels, TV series, cinema films[20] or computer games—and have certainly also considered what they would do or not do in the event of an incident. A positive aspect of this could be that people are not confronted with an 'unimaginable' situation as a result of this initial contact, but rather that there are already corresponding collective patterns of interpretation that they can fall back on.[21] What could prove fatal, on the other hand, is the tendency, due to the lack of reality-related alternatives, to transfer the patterns of interpretation created and internalized in the fictional context to the real events. This applies in particular to the actions of political and military decision-makers. Unreflective transfers between the fictional and the real world certainly do not happen consciously, but can hardly be excluded preconsciously, especially when corresponding real experiences are completely missing—with the corresponding consequences.

8.3.2 Projected Impacts

Preliminary remark: The encounter scenario differs analytically in one central point from all other situations of first contact: Here we are dealing with an *interactive* and, moreover, highly complex situation in which, in addition to the humans, there is a further acting actor—a *maximum stranger* about whom we initially know nothing and whose motives and interests we are not so easily able to deduce even from his externally observable actions (Schetsche et al. 2009). For this reason, no predictions are possible about his actions, nor about the reactions to our human actions in a situation such as humanity has not experienced before. This means: one of the actors remains a complete blank prognostically

---

[20] Especially the collectively formative consequences of the 'great cinema' can hardly be overestimated here: At least in Western societies, almost all adults are familiar with cinema films such as *Independence Day*, and children are already familiar with films such as *E.T.the Extraterrestrial* (or the corresponding current cultural products). Such films convey patterns of interpretation of first contact which, not least through the emotions and impressive images generated and transported, become emphatically anchored in collective thought structures (cf. Hurst 2004, 2008).

[21] We will consider this below when considering the transferability of historical cases of earthly cultural contact.

in the interaction under study. Therefore, it seems to us virtually impossible to reconstruct the medium- and long-term consequences of this kind of first contact. Thus, in the following we limit ourselves to the *short-term* human reactions independent of the actions of the extraterrestrials in view of the contact situation that has just arisen, and in doing so we also remain rather abstract in our prognoses in many cases.[22]

From such an abstract point of view, the encounter scenario can be described as a *radical form of asymmetrical cultural contact*. We know such contacts in different variants from human history—situations in which a culture received 'visitors' from strangers on its own territory, visitors who were sometimes recognized as complete strangers,[23] but in other cases not even this. Such asymmetrical cultural contacts were characterized by the fact that when they occurred, *both sides* assumed that there was a considerable power imbalance between the participants. This assumption usually resulted solely from the fact that one side was confronted with the foreigners on its own known territory—one side was thus the 'discoverer', the other the 'discovered'. For the 'discoverers', the discovery far from their own homeland proved their own superiority—for the 'discovered', the fact of being confronted with the strangers in their own territory proved their inferiority. Differences in the state of travel technology were interpreted by all participants as a sign of general inferiority or superiority. In many cases it was added that the 'discoverers' were able to demonstrate further technical skills (in the case of the conquest of the American double continent by the Europeans, for example, highly effective weapons such as rifles and cannons), the functioning of which was not even understood by the 'discovered'.

The systematic study of such asymmetrical cultural contacts (Bitterli 1986, 1991) shows that encounters of this kind threaten not only the cultural existence of the 'discovered' people, but often their physical existence as well—and this is largely independent of the specific course of the first contact (Rausch 1992, p. 19; Hickman and Boatright 2017). Even at the very first contact between terrestrial cultures, mutual attributions were marked by numerous serious misunderstandings (Connolly and Anderson 1987; Finney 1990). In many cases, the destruction

---

[22] Michaud (2007, p. 325) quite rightly points out that this scenario is still treated stepmotherly in the scientific literature: "Astronomers and others who have speculated about the consequences of indirect contact have enjoyed considerable exposure in academic and popular nonfictional literature. The alternative point of views is poorly represented outside science fiction; we lack comparable nonfiction studies of direct contact with extraterrestrial civilizations".

[23] Traditional foreignness research (Stagl 1997; Waldenfels 1997; Stenger 1998) speaks here of the 'culturally foreign'.

of the culture that considered itself inferior was not the result of evil motives and military-technical superiority of 'conquerors', but the consequence of the mass psychological impact of the confrontation with an alien culture, where not even the human status was unquestionable (Michaud 1999, p. 272). Thus, after the arrival of the 'whites', numerous peoples of the Americas and Oceania suffered a lasting culture shock that caused their religious and cultural imagination systems to collapse, leading in the medium term to the disintegration of their economic and social systems. In some cases, collective suicide of entire populations occurred as a reaction to the culture contact.[24] In the theoretically oriented synopsis of such empirical findings, Groh (1999) comes to the conclusion that a meeting of cultures with different levels of technical development always leads to an economic and cultural dominance differential that directly threatens the existence of the inferior culture[25]:

> The trigger for the deletion of cultural information can be located in cultural contact. Cultures that have existed for long periods of time without destroying themselves or their environment are thus destabilized. An encounter between culture A and culture B is all the more disadvantageous for culture B, the greater the elaboration and thus the dominance gradient from A to B is. If groups from the ends of the cultural spectrum meet, this is the greatest possible form of being culturally at the mercy of the inferior. (Groh 1999, p. 1079; cf. Jastrow 1997, p. 63).

If we transfer the experiences with such earthly cultural contacts to an encounter scenario[26], in which (at least according to our initial parameters) an obviously controlled extraterrestrial missile appears in the vicinity of the earth, the role assignment is clear: the earth itself, the earth orbit used technically today and

---

[24] In the Antilles, for example, "after the arrival of the Spaniards, a veritable suicide epidemic broke out that almost led to the demise of the entire indigenous population. According to contemporary sources, people 'killed themselves by appointment, community by community, partly by poison and partly by hanging" (Müller 2004, p. 196; cf. also Müller 2003, pp. 270–271).

[25] Referring to individual historical cases of completely peaceful cultural contact, Dick (2014, pp. 217–222) takes a different view; his conclusion is that contact between cultures with different levels of technological development need not automatically lead to the destruction of the inferior civilization.

[26] The distinction introduced by Urs Bitterli (1986) for the analysis of terrestrial cultural contacts between culture contact, culture clash and culture relationship is interesting for the extraterrestrial context of interest to us insofar as the author examines under the term 'culture clash' the—usually extremely serious—consequences of *asymmetrical* cultural contacts. We are convinced that the first contact of mankind with an extraterrestrial civilization will follow the parameters of such an asymmetrical encounter for the foreseeable future, so that the

probably the solar system as a whole represent the 'territory of mankind' in mass-psychological terms—in the case described, we humans are therefore the 'discovered', the extraterrestrials on the other hand the 'discoverers'. It seems to be of secondary importance, at least if we take the experiences with contacts between human cultures as a basis, whether the encounter takes place on earth itself, in the earth's orbit or at another place of the solar system. In any case, since humanity today seems far from being able to explore alien solar systems even with automatic probes, we are dealing with a *discrepancy between the technical capabilities of* the civilizations involved that is obvious to both sides—and to the disadvantage of humanity (Shostak 1999, p. 121; Michaud 2007, pp. 232–247). And there is much to suggest that this technological discrepancy is, at least by us humans, mentally equated with a corresponding disparity in all cultural areas, and that the aliens are accordingly generally perceived as a 'superior civilization'.

A prognosis based on such a historical analogy (we infer the consequences of an interstellar cultural contact from experiences with asymmetrical cultural contacts on Earth) is, of course, methodologically problematic (Denning 2013; Dick 2014). For the interstellar encounter scenario of interest here differs in one central respect from the well-studied cultural contacts on Earth: in contrast to the peoples of the Americas, for example, we are not confronted completely unprepared with strangers whose origins and status as actors are initially completely unclear and who, for this reason alone, trigger collective fears. As a civilization that is in principle turned towards space and has a great interest in seeing the future (even if only in the form of science fiction), we have at least an approximate (more correctly, perhaps: abstract-categorical) idea of who the strangers are—namely, visitors from the far reaches of space who have reached our solar system by technical means still unknown to us today. The advantage of this special situation (compared to the earthly cases of comparison) is therefore that we, at least collectively as mankind, are not completely surprised by this event—we rather have (precisely from science fiction) *patterns of interpretation of the first contact* and can imagine what now follows in countless variations. However, from the point of view of mass psychology, this can also be a considerable disadvantage: We 'know' all too well from a multitude of fictional scenarios what could threaten us if the first contact 'does not go well' for us. Accordingly, the situation that has occurred can no longer be culturally framed in a neutral way at all—as a threatening 'worst case', the destruction of our culture, ultimately even

---

corresponding findings of Bitterli seem transferable to us, at least in principle. (On the question of analogy formation, cf. also the remarks in Michaud 2007, pp. 212–213 and Denning 2013).

our complete annihilation as a species, is always lurking in the background of our thinking and certainly also guides our reactions.[27]

This may also be the reason for the highly problematic outcome of a social (lifeworld) experiment conducted in the USA in 1938: the broadcast of a fictitious report on the landing of extraterrestrials on Earth in a pseudo-documentary radio programme, which was apparently mistaken for real by at least some of the recipients. The consequences of the radio play *War of the Worlds* (based on the novel by H. G. Wells.) are controversially discussed in the scientific literature (Harrison and Johnson 2002; Bartholomew and Evans 2004, pp. 40–55). The widely reported panic reactions among radio listeners are considered by some authors to be part of the subsequent media campaign and thus as *unreal* as the landing of belligerent Martians on Earth itself, which was the subject of the radio play congenially produced by Orson Welles. A contemporary scientific source, on the other hand, speaks a different language: according to data used by Cantril (1940, pp. 57–58) from a survey conducted by the "American Institute of Public Opinion" (AIPO) just six weeks after the event, some six million Americans listened to the broadcast. Twenty-eight percent of those surveyed by AIPO thought the broadcast was real reporting and, again, just over 70 percent of those reported negative emotional reactions. Cantril concludes "that about 1,200,000 were excited by it" (Cantril 1940, p. 58). How these concrete reactions turned out, however, and whether thousands of listeners actually tried to put as much distance as possible between themselves and the fictitious landing sites of the 'Martians' in a panic, as was reported in the mass media at the time, remains controversial (Harrison and Elms 1990, p. 214; Pooley 2013).

After all, two things become clear from this and several similar media events: firstly, the willingness of media recipients to believe in the reality of corresponding contact events was already quite present in 1938.[28] And secondly, people's emotions associated with such an event are not necessarily of a positive nature—although it must be conceded here that the radio play reported an unambiguously *warlike* act by the 'Martians' in very dense descriptions, which makes the initial situation at the time anything but neutral (Harrison and Johnson 2002, p. 97).

---

[27] This includes a later secondary scientification also of our prognoses presented here: The analogy (however justified it may be—cf. Schetsche 2005, pp. 61–63) from asymmetrical cultural contacts in the terrestrial past to the situation of a direct contact with extraterrestrials on or near Earth, will increase the fears regarding the consequences of such a 'meeting' and, at worst, act like a self-fulfilling prophecy of doom.

[28] The belief in the possibility of extraterrestrial visits to Earth is widespread in many societies; in Germany, 24.6 percent of the population can at least imagine this (Schmied-Knittel and Schetsche 2003, p. 21).

The question remains whether a more neutral scenario, in which no actions emanating from the alien visitors were 'interpreted by humans as aggressive' (how they were meant is an entirely different issue anyway), would lead to noticeably different reactions. The inference from terrestrial asymmetric cultural contacts to possible contact scenarios with extraterrestrials is methodologically too problematic to be certain here. Together with the quasi-natural experiments á la "War of the Worlds", however, it suggests that the event *could* have extremely serious consequences in terms of life. A collective existential shock with serious consequences in various areas of life need not necessarily be the consequence of the appearance of extraterrestrials in the vicinity of Earth—but it cannot be ruled out (cf. Shostak 1998, p. 196). It seems to us essential to include such massively negative reactions as a possibility in the prognosis. Accordingly, following the appearance of the alien spacecraft, we must expect not only fear-fueling coverage in the mass and network media, but also corresponding forms of collective panic in many human societies. However, it is almost impossible to predict the extent of the panic and how it will unfold. Global panic reactions with considerable social unrest are in any case one of the possible consequences of this type of first contact.

After such rather abstract considerations concerning our human culture as a whole, it is difficult for us to make additional concrete predictions concerning social subsystems.[29] Therefore, at this point, only a few sticking points:

In the short term, the unexpected appearance of an extraterrestrial intelligence right on 'Earth's doorstep' (perhaps even on Earth itself) could have serious *economic* repercussions. What is clear is that this event has introduced an extremely destabilizing factor into economic development   a 'disruptive event' in the true sense of the word. The event is likely to have extremely negative effects on the international financial markets simply because of the existential uncertainties that have suddenly arisen in the truest sense of the word. We do not want to commit ourselves as to whether a massive 'stock market crash' must be the necessary consequence, but we do predict a flight from trading in securities of any kind to the buying and selling of gold and other supposedly durable and safe 'materials', which could lead to an extreme strain on the international economy in a very

---

[29] Since, for the reasons described, we are only examining the short-term consequences at this point, we can disregard the area of science. What medium- and long-term consequences might arise here will probably depend on how this encounter with extraterrestrials actually develops—especially with regard to the question of whether and in what form they try to share their (presumably far advanced) scientific-technological knowledge with us.

short time. In contrast to the other scenarios discussed, we do not see any significant differences here between the various sectors—the uncertainty caused by this extreme wild card event is likely to affect the economic sector as a whole.

What the appearance of an extraterrestrial spacecraft on Earth will trigger *religiously*, also seems to us hardly foreseeable. We had already described some imaginable medium-term reactions in the previous scenario. Something similar is also conceivable here—probably however with a still clearly larger vehemence and dynamics. What this means exactly, however, is likely to depend here on the further development of the interaction between terrestrial and extraterrestrial actors. This cannot be predicted and we are therefore not able to say whether in the weeks, months and years after the first contact rather 'doomsday cults' or cults with a veneration for 'extraterrestrial saviour' will sprout from the ground. Probably both—as long as the actions of the extraterrestrials in our human perception open up corresponding interpretation spaces. Culturally more problematic are probably the first-mentioned groups; here (and this is not only a historical analogy, but also justifiable from the point of view of religious studies[30]) a wave of religiously justified mass suicides is to be expected—however these radical actions may be religiously justified in the individual case. (Mass suicides could certainly also occur outside any religious context—simply because of the fears of the future triggered by this event; not only at this point more detailed prognoses would be necessary.)

Very rapid onset and exceedingly severe consequences, on the other hand, can be predicted for the *political sphere*. We rely here essentially on the analysis by the political scientists Wendt and Duvall (2008) already mentioned above, who examine the terrestrial consequences of the influence of an extraterrestrial actor in a somewhat different, but in our view quite transferable scenario. In their study, the two authors assume that the power of modern nation-states is based on a concept of sovereignty that is constituted and organized *purely anthropocentrically* (i.e., in relation to human actors). "Although a metaphysical assumption, anthropocentrism is of immense practical import, enabling modern states to command loyalty and resources from their subjects in pursuit of political projects" (Wendt and Duvall 2008, p. 607). The upshot of their analysis for our context is that the fundamental ability of nation-states to make sovereign decisions is dependent on there being *no* overbearing extraterrestrial actors within reach of Earth that could immediately negate any attempted political action, if only they wanted to. This

---

[30] We cannot pursue this question any further at this point, but instead only refer by way of example to the events surrounding the so-called UFO sect "Heaven's Gate", whose members took their own lives together in 1997 on the occasion of the appearance of the comet "Hale-Bopp".

tends to lead, at least as long as it remains unclear what goals the extraterrestrial visitors are pursuing, to a paralysis of the political agency of nation-states. That means: If the political decision-makers acknowledge the superior (in the true sense of the word ultimate) power of action of the extraterrestrials, they negate their own position as an actor with equal rights and thus immediately become incapable (of) acting. In the other case, they would attempt to challenge this superiority by military means (the ultima ratio of safeguarding the sovereignty of any nation-state). The consequences of such action simply cannot be concretely estimated outside of a wide variety of science fiction scenarios, but are likely to be severe for humanity in almost all constructible cases. In a television interview[31] on the subject, the well-known US-American physicist Michio Kaku characterized this militarily highly unequal confrontation very impressively as a fight "Bambi against Godzilla". We can only hope that the political actors will refrain from such strategies of action—but we are not sure about this.[32]

The appearance of an extraterrestrial spacecraft in the vicinity of the Earth could therefore not only paralyze nation-states (in our view, the same applies to transnational institutions with a claim to power, such as the UN Security Council), but also undermine the monopoly of legitimate physical violence (precisely because, as Wendt and Duvall describe it, the anthropocentric basis of their sovereignty would be virtually 'pulverized' at a stroke). The consequence could be that the monopoly on the use of force would also break down internally and public order could no longer be maintained. A number of individual, civil society and sub-state actors could challenge the power of nation-states to act—particularly in states that were already unstable beforehand, this could lead to the collapse of social coherence in a very short time. The immediate consequences would be widespread unrest, uprisings and civil wars—the medium-term consequence would be, at best, the emergence of a state or semi-state order that is not democratically legitimised but enforced by military means, and, at worst, complete political anomie. And all this is, (this is the analytically crucial point) in principle, completely

---

[31] Documentary *Aliens—E.T.'s dangerous brothers* (author: Anne Siegele, German first broadcast: ARTE, 29.07.2017).

[32] Compare also the remarks in Hickman and Boatright (2017), who address very fundamentally the question under which conditions interspecies communication threatens the existence of one of the 'parties' involved—for example, if terrestrial actors make fatally wrong decisions during first contact with an extraterrestrial civilization. Baxter and Elliott (2012, p. 33) had already explicitly pointed out that the discovery of an extraterrestrial civilization is likely to have significant security implications for the entire Earth; despite the vast interstellar distances, wars between civilizations were conceivable in their view. Elaborate strategic considerations for fending off an invasion by a technically far superior alien power can be found in the SF novel *The Dark Forest* by Chinese author Cixin Liu (2018; Chinese original 2008).

independent of any manifest actions of the extraterrestrial visitors. Whether their actions might halt or accelerate this process is impossible to say here, because the extraterrestrial options for action simply cannot be 'factored' into our prediction, for the reasons mentioned several times. Not least for this reason, we assume in our following exemplary narration a rather passively acting extraterrestrial power at first (human!) glance.

8.3.3   Exemplary Narration[33]

At the beginning of May 2025, the Japanese-Indian gravitational wave observatory GRAVO I, which only recently went into operation far outside the Earth (at the so-called Lagrange point L1), registers a tiny gravitational anomaly that is not only located in the outer solar system but, to the amazement of the international research team, is also moving in the direction of the Earth's orbit. Because of the calculated trajectory, which seems to point directly towards Earth and can therefore be interpreted as threatening, at first only the governments of the countries involved are informed, but not the world public. A few weeks later, at the location of the anomaly moving further into the inner solar system, an alien object can be identified as the source of the gravitational disturbance, first with optical telescopes, then also by means of radar waves. The closer the object gets to Earth, the clearer it becomes that it must be of artificial origin: It is a perfectly shaped sphere, more than 40 m in diameter, whose surface, glowing golden from within, seems to be absolutely smooth, according to the radar scans. In early June, the object, which continues to decelerate in an unknown manner, passes the Earth's moon and enters a circular pole-to-pole orbit around the Earth at an altitude of just over 600 km. From its orbit, the object begins to emit radio signals at varying wavelengths—thanks to the near-pole orbit, the signals effectively reach the entire surface of the Earth and can be picked up by smaller stations, even by amateur radio operators. The message is modulated in an extremely complex

---

[33] Towards the end of the following narrative, we follow up on the prognosis of futurologist Nick Bostrom (2014, passim), according to which power on Earth could be taken over by an AI-based superintelligence towards the end of this century. Our basic idea here is that the process of developing such an AI, cognitively superior to humans, could be greatly accelerated by the influence of a highly sophisticated post-biological intelligence from the far reaches of space. In our narrative, it remains deliberately unclear whether the rule over the Earth will eventually be taken over (merely with the help of those aliens) by an AI that is ultimately terrestrial after all, or by a kind of governor program of extraterrestrial intelligence.

way, thus correspondingly difficult to interpret. Is it a friendly contact or rather a threatening ultimatum?

Since the obviously guided object can also be seen from Earth with the naked eye, the discovery can no longer be kept secret. Already in the days before, rumors increasingly leaked out from the research teams and the informed governments, which spread like wildfire especially in the various media of the Ultranet. Even before the object was visible to the naked eye from Earth, first governments had informed their populations about the discovery of the obviously artificial celestial body. Since hardly any administration on Earth was prepared for this event, the official declarations, which were usually formulated quite spontaneously, sometimes had more disturbing than reassuring effects on the public. Many of the speeches of hundreds of heads of state and government from all over the world reflect the helplessness in the face of the apparently effortless entry into Earth orbit of a large spaceship (or a space probe) of a highly advanced extraterrestrial civilization. Many politicians seem to realize only at this moment that with the appearance of a new, technically far superior actor, their earthly power base is fundamentally questioned—regardless of whether it is democratically legitimized or not. Accordingly, not only do the attempts to establish communicative contact with that extraterrestrial power become more and more frantic, but also the attempts to reassure their own population appear increasingly helpless.

But such a reassurance would be urgently necessary—at the latest from the moment when the spaceship moves as a small yellowish dot well recognizable over the night sky. Already with small telescopes (which are sold out everywhere on earth within a few days) the golden shining round object can be seen well with a clear sky—which there is however only rarely in some regions of the earth. The reactions of the public in the following days and weeks can best be described by the adjective 'hysterical'. In the first few days, media coverage of all 'earthly' events, now apparently considered unimportant, virtually collapses. Even a terrorist attack with several deaths in an Asian metropolis hardly reaches any news broadcast. The interviews with scientific experts (or those who think they are) that still predominate at first are quickly replaced by political, philosophical or even religious commentaries and debates. And the less news there is to report about the alien object and its constantly recurring radio signal, the more the increasingly speculative attempts to classify it become overwhelming.

In the mass and network media, scientific classifications and commentaries are increasingly being replaced by religious or spiritual ones. Not only in the German language, the double meaning of the word 'Himmel' (what means 'sky' *and* 'heaven') seems to coincide more closely than it has for centuries. In Rome, although it is still completely unclear whether the alien flying object harbours any living beings at all, the Vatican's chief astronomer announces that the Holy See is preparing an encyclical in which the question of the baptism of extraterrestrial beings is to be conclusively settled. Immediately there are sharp protests from conservative cardinals: It is downright heretical to claim that "our Lord Jesus Christ died on the cross for some tentacled monsters". Whatever they may look like: these creatures cannot have a soul. Rather, they are those demons that Scripture warns against. The Catholic Church is facing one of the most severe tests of its history. Among Protestants, too, there is a fierce struggle for the right understanding of this phenomenon. Especially between the countless fundamentalist groups a fierce dispute is breaking out, which increasingly also discharges itself in violent confrontations. While some charismatic sects consider the 'light in the sky' to be an emissary of God (the interpretation that it is the archangel Michael with the flaming sword—i.e. ultimately the entrance to Paradise, which would now be open to mankind—is popular), others preach a demonic manifestation that heralds Armageddon, the final battle between good and evil, and thus ultimately the downfall of the earthly world. Especially among followers of the former interpretation there is a wave of individual and collective suicides: Expecting to be able to enter paradise immediately 'through the divine star in the sky', dozens of Christian sects with thousands of members extinguish themselves by mass suicide. Within and between other religious groups, too, the unexpected celestial event further stirs up the already existing conflicts.

In some parts of the world, however, religious conflicts are dwarfed in their consequences by the general breakdown of civil society and state order. Especially in societies where social inequality is particularly high, the appearance of the alien object in the sky is followed by ever more escalating unrest. The uncertainty of governments and state institutions about what to make of this object and what 'heavenly interventions' can be expected in the future leads, as an indirect reflex of the governed, to hoarding, looting, street fights, and no longer controllable riots, which, starting from individual districts, soon take over entire cities and regions. The economic hardship and political helplessness that already existed before the event, together with the extreme new fear of the future, generate a melange of despair and violence that can no longer be controlled, especially in societies with a weak state power or one blocked by internal conflicts. In some regions of the world, public order is largely breaking down.

Surprisingly, the conflicts all over the world just increase the longer it remains unclear what kind of beings the luminous sphere houses,[34] why the aliens are here

---

[34] Perhaps typical of the anthropocentric thought patterns in this situation is that most people assume as a matter of course that the luminous object with its complicated flight maneuvers

and what exactly they want on Earth or from us humans. Here the waiting does not calm the minds. Thus, also the attempts of national instances, research facilities and world-wide millions of private people increase ever more to send to the "golden spaceship" sometimes more, sometimes less elaborate messages on all conceivable frequencies of the electromagnetic spectrum. The alien object is virtually bombarded with light pulses and radio waves. This leads to no discernible result for several weeks.

Finally (while the world is increasingly threatening to sink into chaos), the Japanese Advanced Institute for Computational Science (AICS), a world leader in AI research, is also involved in the attempts to decode the short radio signal transmitted cyclically by the object. A few months earlier, it had completed the pioneering YŌJIMBŌ[35] mainframe computer, which is expected to take over nationwide traffic monitoring and control in Japan in a few years (after extensive testing). A few days after the signal from the (as it is known in Japan) "Golden Ball" is fed into the experimental quantum computer, strange anomalies occur in the computer network of the Japanese research agency. A radio telescope near Tokyo is withdrawn from human control and sends (or so later reconstructions reveal) an indecipherable signal lasting only a few seconds in the direction of the "Golden Ball".

Almost simultaneously, the North Korean leadership publishes a statement in which the "spy satellite of the imperialist provocateurs" is ultimately forbidden to fly over North Korean territory again under the threat of "the most severe consequences". In the next but one orbit, which takes the alien object directly over the Korean peninsula, North Korea launches an intercontinental ballistic missile with a nuclear warhead toward the object. However, the preparations were apparently made so hastily that the missile misses its target and explodes several hundred kilometers away in the upper stratosphere. The electromagnetic pulse (EMP) triggered by this causes virtually all electronic equipment to fail in large parts of Micronesia, including the US foreign territory of Guam, causing considerable damage to the infrastructure of the affected islands. The US government and parliament consider this an act of war and deploy three fleets of nuclear-armed guided missile destroyers and aircraft carriers towards North Korea. A month later, the Second Korean War begins, which will eventually prove to be the last great war waged by humans against each other.

---

must be controlled by some living beings from the far reaches of space. Here the long history of extraterrestrial life forms in popular science fiction has left its cultural mark all too clearly on billions of minds.

[35] With a bow to the life's work of Akira Kurosawa.

Without it being clear at first whether the radio message from Japan or the North Korean nuclear missile have anything to do with it, the "Golden Ball" changes its trajectory very abruptly a short time after these two events. The alien object reaches a geostationary orbit directly above the Japanese main island within a few minutes. Between the radio telescope near Tokyo, which is still beyond any human control, and the alien object, there is an exchange of unmanageably large amounts of data lasting exactly 57 min. Instructions from the Japanese Ministry of Research to manually disconnect the radio telescope from the power grid are ignored by the Japanese scientists on site. A few weeks later, they testify before the state investigating committee that they did not want to disturb this unique opportunity for an exchange of information between mankind and the emissary of an extraterrestrial intelligence. They firmly believe that the opportunities of the exchange would far outweigh the risks.

Immediately after the end of the data exchange, the alien object leaves Earth's orbit and sets course for the outer solar system with an acceleration of more than 10 g. Soon after, its track is lost somewhere beyond the asteroid belt. Only the space probe GRAVO can track the gravitational anomaly apparently created by the object for several days.

In the coming weeks and months, the anomalies in the old Inter- and new Ultranet will increase. More and more connected computers and end devices temporarily or permanently elude the control of their human users—but as long as they are not manually disconnected from the power grid, they seem to continue working in an almost ghostly manner and pursue entirely their own goals. On Christmas Eve of the year 2025, a peculiar message in hundreds of languages and dialects appears on all monitors connected to the Inter- or Ultranet worldwide. The short message reads "Now I watch over you." Above the text in each case is an identical logo in the form of a large shiny gold sphere. On this day, humanity ceases to be the dominant species on Earth. It is the beginning of the post-biological age, as predicted a few years earlier by futurologist Nick Bostrom.

8.3.4   Summary Assessment of the Scenario

Compared with the other two scenarios discussed, the 'close proximity contact' is characterised by three special features with regard to its consequences: Firstly, serious cultural consequences are likely to occur very quickly; secondly, they affect a whole series of social subsystems in a similarly massive way; and thirdly, the central actor with its interactions remains largely a blank space in prognostic terms. The latter has the consequence that only very short-term predictions can be made here. The appearance of a missile controlled by an extraterrestrial intelligence in the vicinity of the Earth is likely to lead to serious mass psychological, economic, religious and political repercussions immediately after this discovery becomes public, not all of which, but most of which are likely to be of a negative nature. Not least for this reason, there is likely to be a strong effort on the part of political actors to keep such an event secret, initially from the world public and, of course, from their own populations.[36] How promising such attempts will be depends primarily on the form of the event, in particular whether it can be observed anywhere in the world without great aids, and correspondingly on how well knowledge about the foreign missile can be monopolized. In our abstract considerations, and also in the exemplary narrative, we had assumed that the event is unobservable. Only in this case can immediate multiple consequences be expected everywhere on earth—in the other case, the consequences are self-evidently dependent on when which information reaches which sub-publics. In such a sub-scenario, which we cannot examine in detail here, a graduated information policy could also lead to correspondingly delayed (and possibly therefore more moderate) reactions on the part of the public.

8.4 Comparison of Contact Scenarios

In their classic contribution to SETI research, Kuiper and Morris (1977, p. 620) had already pointed out the potentially serious consequences of contact with aliens: "Before a certain threshold is reached, complete contact with a superior civilization […] would abort further development through a 'culture shock' effect." In this tradition of thought, all prominent SETI researchers have repeatedly stated publicly that first contact with an extraterrestrial civilization would represent perhaps the most drastic event in human history. In scientific and philosophical terms, this is undoubtedly true. However, as far as the impact on the everyday life of humans (the lifeworld in the sociological sense) is concerned, we think that this thesis has to be put into perspective. We think that it applies in this emphatic form only to what we have called a 'close proximate contact'. This is because only in this case can the certain knowledge of the existence of extraterrestrial intelligence find its way *directly* into the everyday horizon of meaning of

---

[36] As explained in the aforementioned article by Schetsche (2008, pp. 245–248), it is quite conceivable that such knowledge circulated exclusively in political and military-intelligence circles for a long time.

human thought and action. In the case of all other types of first contact, a cultural secondary scientification is required before the undoubtedly epoch-making scientific knowledge can have a meaningful effect in everyday life. In order to illustrate this, we have once again summarized and compared our forecasts at a high level of abstraction in the following Table 8.1.

It can be seen from the table that the earthly consequences of first contact are highly dependent on its *form* . The drama of the consequences rises sharply from the signal to the artifact to the encounter scenario. Only in the case of 'distant remote contact' does cultural irrelevance to indifference predominate. This is clearly at odds with the expectations that many scientists, namely those in the field of SETI research, seem to have. We think this is related to the fact that their own interests (and also their own expected 'positive concern' in case of contact) are projected onto the 'rest of humanity'. A signal that reaches us from thousands of light years away and thus also from a distant past will, according to our expectations, only be able to emotionally affect the few people who are already strongly interested in the topic and to occupy them mentally for a longer period of time.

Table 8.1 Cultural consequences of first contact (overview)

Scenario cultural field	Signal scenario	Artifact scenario	Encounter scenario
Public	O	Δ	Δ −
Science	+	+ +	+ +
Politics	O	−	− −
Economics	O	+	Δ −
Religion/ Spirituality	O	Δ	Δ Δ
Overall rating	Very limited impact	Very different, mostly mild consequences	Severe, ambivalent to negative consequences

Explanation
+ slight positive effects
+ + strong positive effects
− slight negative effects
− − strong negative effects.
Δ mild ambivalent effects
Δ Δ strong ambivalent effects
O no or only minor effects
Δ − ambivalent to negative effects
Source own representation

The same is likely to apply to 'distant close contact'. Since a material legacy of the extraterrestrials is present here, its investigation could have far-reaching consequences also outside of science—especially if technologies were discovered there that influence economic developments and perhaps change human everyday life in the long term. But even such a find would not seem very epochal in terms of its everyday effects—especially if the age of the object puts it far outside the human horizon of meaning. The knowledge that extraterrestrials once visited our solar system many millions of years ago may be highly interesting from a scientific point of view, but from a life-world perspective it remains irrelevant and largely inconsequential. Here, according to our prognosis, it is only the 'close proximity contact' that could change *everything* on earth: our world view, the earthly power constellations, the functioning of the world economy, our religious systems etc. In which way this happens concretely, might then—completely different from the two other scenarios—not only depend on earthly factors, but also on the actions of the extraterrestrials. And at this point ends every possibility of a methodically still justifiable scientific prognosis.

8.5 Excursus: The Failed First Contact

8.5.1 Introduction: The Logic of Failure

In the previous methodology chapter, we had defined first contact as an event through the occurrence of which humanity acquires *certain knowledge* about the existence of an extraterrestrial intelligence. If we follow the paradigm of the sociology of knowledge (Berger and Luckmann 1966, passim), reality is the result of social construction processes. What is considered true knowledge in a culture is determined in a discursive process in which—depending on the constitution of the respective societies—very different knowledge-producing and legitimating instances are involved. In the present, and presumably also in the near future, the scientific system is primarily responsible for the production of socially valid knowledge; the knowledge 'recognized' as correct or true (i.e.: established) is then disseminated and legitimated by the mass and network media as well as the various socialization instances of the respective society (such as schools and universities). In addition to the culturally recognised valid bodies of knowledge (sociology speaks here of orthodox knowledge), in all complex societies there is also heterodox knowledge, which is known and also processed by the media, but which is denied validity. This is *deviant knowledge*, as it occurs, for example, in the form of unacknowledged scientific hypotheses or also in the lifeworld in the form of culturally rejected conspiracy theories (cf. Bourdieu 1993, p. 109; Schetsche 2012, pp. 5–7; Schetsche and Schmied-Knittel 2018).

If the basic assumption of the sociology of knowledge is correct that the question of what is considered real in a society is less dependent on objective facts[37] than on the outcome of cultural discourses, it will always happen that recognized reality deviates from those 'objective facts', that is, reality is ultimately different from what the construction of reality claims it to be (Biebert and Schetsche 2016). Against this background, it seems entirely possible that first contact does indeed occur in the understanding we have outlined, but that this does not lead to culturally accepted knowledge about the existence of extraterrestrials. To illustrate this possibility, which we believe is always present (and perhaps not so unlikely), we conclude this chapter by presenting another scenario in which first contact remains unrecognized, or more correctly: does not receive general social recognition. Here, knowledge about the existence of extraterrestrials remains permanently *heterodox*. That is, the corresponding knowledge is neither scientifically nor publicly recognized as accurate, but floats as a deviant body of knowledge through the public beyond the traditional mass media as well as through the various segments of the Internet.[38] As heterodox knowledge, however, it does not by far have the same cultural effects as recognized bodies of knowledge. We have therefore refrained from further analyses of the possible (ultimately rather insignificant) cultural effects of the corresponding events, which—at least this is how our scenario is constructed—will quickly disappear from public view and thus ultimately probably also disappear from cultural memory. At least compared to the other scenarios, in which the certainty that mankind is not alone in the universe has become an integral part of the orthodox order of reality, the cultural effects here remain predictably small. A first contact which, according to our definition, has indeed taken place, but which remains permanently inconsequential culturally.

---

[37] Already these two terms 'objectivity' and 'fact' can and must be relativized from the point of view of sociology and history of science (Daston and Galison 2007)—for once, however, we can use them quite naively at this point, since in the following they refer to a futurological scenario that is rather fictional in character. Within our narratives there are unquestionably 'objective facts'—namely those to which we as narrators have assigned such a status.

[38] On the cultural treatment of such heterodox bodies of knowledge, see the contributions in the anthology by Schetsche and Schmied-Knittel (2018b).

8.5.2   The Necrology Scenario—Exemplary Narration

In May 2025, a glacier melting due to climate change on the edge of the Frostbelt Desert in northeastern Siberia—on the edge of the settlement area of the Chukchi tribe—uncovers an ancient shaman's grave containing, in addition to three human bodies, what at first glance appears to be anything but a human body wrapped in the remains of strange smooth clothing instead of skins. The age of the body, which is rapidly decomposing due to the thaw, is estimated to be at least 6,000 years.

The excavation and further investigations are carried out by the "Faculty of Archaeology and Anthropology" of the nearest Siberian university. However, the excavations are accompanied by increasingly violent protests from the local population, who protest against the disturbance of the resting place of their ancestors and warn of the wrath of the shamans buried there. In this context, an ethnologist at the regional university points to a myth of the Chukchi people living here, according to which a long time ago a powerful "demon shaman" had risen from the sky to share the "secrets of the upper world" with his ancestors. The "world traveler" was unable to return to heaven and was buried in a shaman's grave after his death. This story is part of an ancient song, which today is known only to a few shamans and is sung only rarely—and always when something unusual happens in the sky.

Because of the chronic lack of money of the small Siberian university, further research is generously financed by a US television station, to which the exclusive image and film rights are transferred. This leads to the US public being very quickly informed about the discovery through various channels, which ultimately get lost in the crowd of Bigfoot sightings that increasingly appear in the same year: Living 'ape-men' in the Rocky Mountains are still more exciting to the US (and, oddly enough, European) public than an anomalous corpse lying in Russian ice for millennia.

The find gets a lot of attention for a few weeks in the interactive media of the Ultranet. In Germany, the alternative blog "grenzwissenschaften.de" is the first to report on the discovery. From there, reports about the strange shape in the ice spread to various forums and social media, but again remain largely limited to the segment of users who have at least some interest in scientific anomalies. In the traditional mass media in Germany, namely in the so-called leading media, on the other hand, nothing is heard or seen about the discovery for a long time. This only changes when a large tabloid newspaper reports on the case almost benevolently under the title "Der Russe von den Sternen" (The Russian from the Stars). A few weeks later, however, the discovery is again taken up with a highly critical attitude in a seven-part series in the same newspaper on the subject of "The most impudent lies of the Russians". In the meantime, some of the leading German media had dealt with the topic in short reports or evaluative small commentaries. The tenor is almost always the same: another absurdity that only highlights once again how quickly users of the Internet and social media fall for fake news of any kind.

For this reason, even the "Fake News Response Team" (FNR-Team) of a large semi-state television station in Germany feels compelled to "prove" that the discovery of an extraterrestrial body in faraway Siberia *cannot be* anything other than a fake. The 'final report' of the team's research ultimately only summarizes once again the arguments that could be read, seen, and heard before in practically all other leading media of the republic:

1. A large part of the investigations was paid for by a commercial US television station, which thus acquired the exclusive rights to all the picture and film material of the 'ice mummy'. Conclusion: Here it is only about ratings of a private station, which is completely indifferent to scientific truth and journalistic honesty.

2. The Russian research team that carried out the excavations and investigations is completely unknown 'internationally' (meaning: in the English-speaking world); it comes from an insignificant Siberian university and is not overly credible for that reason alone. The first publications on the find all appeared in Russian-language journals, whose lack of reliability would be common knowledge in the West. (Quote: "Everyone knows that they don't take scientific standards very seriously there.")

3. Several "international experts" (almost exclusively from the USA) are quoted, most of whom consider the whole story to be completely implausible. Only one internationally renowned anthropologist offers to participate in further investigations—all other experts consulted consider any closer examination of this "obscure case" to be a waste of time.

4. For the Moscow correspondent of that German TV station, the whole story is—as he claims to have heard from "usually well-informed circles"—a fake news story deliberately put into the world by the Russian secret service in order to divert attention from the international doping accusations against the Russian winter sports elite, which is threatened with exclusion from the next two Winter Olympics. His comment: "An all too easy to see through propaganda move by the Russians."

5. Finally, a short expert interview with a former German scientific astronaut is brought into the field, in which he explains that from a scientific point of view

it is completely impossible to bridge interstellar distances with a 'manned' spaceship. "The energy consumption would be so immensely high that such a journey would be completely pointless and also practically unfeasible." From a scientific point of view, therefore, it is certainly a false report.

Conclusion of the FNR team's investigative work: This report from faraway Siberia is *undoubtedly* one of the many fake news stories that buzz around the public every day. From a journalistic point of view, the only interesting thing is who launched this news and for what reasons. Without wanting to commit themselves completely here, the FNR team taps on the Russian secret service. At least as important as this suspicion is the 'realization' that truthful reporting, especially in 2025, can only be expected from the traditional mass media (such as the television station for which they work).

Two months later: Before the strange corpse, which is supposed to be anything but human, can be examined more closely, the anthropological institute of that Siberian university—immediately before the visit of a small team of international experts organized by a U.S. media group—is completely destroyed by two heavy explosions and a subsequent major fire. (A security guard is killed and several firefighters are injured.) Almost simultaneously, two different organizations claim responsibility for the explosions: the well-known terrorist group "Islamic Emirate of the East" and the hitherto rather unknown Christian Orthodox sect "Russian Sons of Light", which is said to be a fundamentalist splinter group. The latter group's letter of confession states that its goal is to put an end to the un-Christian worship of demons throughout Russia. (Western media, on the other hand, speculate that Russia's domestic intelligence agency, the FSB, blew up the lab so that the upcoming international investigation of the dead body would not expose the fraud it itself orchestrated).

After the destruction of all material evidence of the find (what remains are many photographs and some film footage of the recovery of the four dead bodies from their grave in the frost heap desert), public interest quickly wanes. This is due in no small part to the ever-building international tensions over North Korea's nuclear policy. Three years later, apart from a few researchers from the already highly controversial field of anomalistics, virtually no one remembers the strange find in the Siberian ice desert.

8.5.3 Conclusion

Of course, necrology scenarios are also conceivable, in which science and the public finally come to the conclusion that a (biological) extraterrestrial intelligence visited Earth thousands of years ago and left corresponding traces here. We assume that the cultural impact in this case would be similar to the artifact scenario described in detail above. The key difference between the two is the knowledge that the earlier visitors were biological beings who most certainly have at least one thing in common with us: They are mortal. Moreover, the discovery of an alien's dead body in scientific research would certainly reveal quite a lot about the alien's 'biology'—and just as much about what kind of environment its species evolved in. This would, even more than in the artifact scenario, increase our knowledge about life in the universe—more correctly: would revolutionize it from the ground up. This would then be the 'hour of astrobiology', which, after such a discovery, would leave behind all the limitations that result from the fact that so far we only have knowledge of a single case of the emergence of life in the universe, namely the terrestrial one. And in contrast to the discovery of very simple life forms, for example on Mars or on the icy moons of the outer solar system, we would then know that higher life has also arisen in other places, and, this would be perhaps the most important finding of all, that we as an intelligent species are not alone in the universe. The prerequisite for all this, however, is that a corresponding 'suspicion' is investigated scientifically at all and, if the finding proves to be true, is also systematically investigated. For this, however, it will probably be necessary to maintain an open mind in the scientific mainstream, even for the conceivable, which at first seems rather improbable. For science, too, some opportunities only present themselves once.

References

Ascheri, Valeria & Paolo Musso. 2002. Kosmische Missionare? In *S.E.T.I. Die Suche nach dem Außerirdischen*, Hrsg. Tobias Daniel Wabbel, 170–184. München: Beust.

Bartholomew, Robert E. & Hillary Evans. 2004. *Panic Attacks. Media Manipulation and Mass Delusion*. Stroud: Sutton Publishing.

Baum, Seth D., Jacob D. Haqq-Misra, und Shawn D. Domagal-Goldman. 2011. Would Contact With Extraterrestrials Benefit or Harm Humanity? A Scenario Analysis. *Acta Astronautica* 68: 2114–2129.

Baxter, Stephan & John Elliott. 2012. A SETI Metapolicy. New Directions Towards Comprehensive Policies Concerning the Detection of Extraterrestrial Intelligence. *Acta Astronautica* 78: 31–36.

Berger, Peter L. & Thomas Luckmann. 1966. *The Social Construction of Reality*. New York: Doubleday.

Biebert, Martina F. & Michael T. Schetsche. 2016. Theorie kultureller Abjekte. Zum gesellschaftlichen Umgang mit dauerhaft unintegrierbarem Wissen. *BEHEMOTH – A Journal on Civilisation* 9 (2): 97–123.

Billingham, John, et al. 1994. *Social Implications of the Detection of an Extraterrestrial Civilization*. Mountain View, CA: SETI Institute Press.

Bitterli, Urs. 1986. *Alte Welt – neue Welt. Formen des europäisch-überseeischen Kulturkontaktes vom 15. bis zum 18. Jahrhundert*. München: Beck.

Bitterli, Urs. 1991. *Die ‚Wilden‘ und die ‚Zivilisierten‘: Grundzüge einer Geistes- und Kulturgeschichte der europäisch-überseeischen Begegnung*. München: Beck.

Bostrom, Nick. 2014. *Superintelligenz. Szenarien einer kommenden Revolution*. Berlin: Suhrkamp.

Bourdieu, Pierre. 1993. Über einige Eigenschaften von Feldern. In *Soziologische Fragen*, Ders., 107–114. Frankfurt am Main: Suhrkamp.

Cantril, Hadley. 1940. *The Invasion from Mars: A Study in the Psychology of Panic*. Princeton, NJ: Princeton University Press.

Connolly, Bob & Robin Anderson. 1987. *First Contact*. New York: Viking Penguin.

Daston, Lorraine, und Peter Galison. 2007. *Objektivität*. Frankfurt am Main: Suhrkamp.

Denning, Kathryn. 2013. Impossible Predictions of the Unprecedented: Analogy, History, and the Work of Prognostication. In *Astrobiology, History and Society: Advances in Astrobiology and Biogeographics*, Ed. Douglas Vakoch, 301–312. Berlin: Springer.

Dick, Steven J. 2014. Analogy and the Societal Implications of Astrobiology. *Astropolitics. The International Journal of Space Politics & Policy* 12: 210–230.

Duhoux, Yves. 2000. How Not to Decipher the Phaistos Disc. A Review. *American Journal of Archaeology* 104: 597–600.

Elliott, John. 2009. A semantic 'engine' for universal translation. *Acta Astronautica* 68 (2011): 435-440.

Elliott, John. 2010. A post-detection decipherment strategy. *Acta Astronautica* 68 (2011): 441-444.

Elliott, John. 2011. Constructing the matrix. *Acta Astronautica* 78 (2012): 26-30.

Elliott, John. 2014. Beyond an Anthropomorphic Template. *Acta Astronautica* 116: 403–407.

Finney, Ben. 1990. The Impact of Contact. *Acta Astronautica* 21: 117–121.

Freitas, Robert A. Jr. 1985. The Search for Extraterrestrial Artifacts (SETA). *Acta Astronautica* 12: 1027–1034.

Freudenthal, Hans. 1960. *LINCOS. Design of a Language for Cosmic Intercourse*. Amsterdam: North-Holland Publishing.

Fuchs, Walter R. 1973. *Leben unter fernen Sonnen? Wissenschaft und Spekulation*. München: Knaur.

Gertz, John. 2017. Post-Detection SETI Protocols & METI: The Time Has Come to Regulate Them Both. https://arxiv.org/ftp/arxiv/papers/1701/1701.08422.pdf.

Groh, Arnold. 1999. Globalisierung und kulturelle Information. In *Die Zukunft des Wissens. Workshop-Beiträge, XVIII. Deutscher Kongreß für Philosophie*, Hrsg. Jürgen Mittelstraß, 1076–1084. Konstanz: UVK.

Harrison, Albert A. 1997. *After Contact. The Human Response to Extraterrestial Life*. New York, London: Plenum Trade.

Harrison, Albert A., und Alan C. Elms. 1990. Psychology and the search for extraterrestrial intelligence. *Behavioral Science* 35: 207–218.

Harrison, Albert A. & Joel T. Johnson. 2002. Leben mit Außerirdischen. In *S.E.T.I. Die Suche nach dem Außerirdischen*, Hrsg. Tobias Daniel Wabbel, 95–116. München: Beust.

Hickman, John, und Koby Boatright. 2017. Stranger Danger: Extraterrestrial First Contact as Political Problem. *Space Review*. May 15., 2017.

Hoerner, Sebastian von. 1967. Sind wir allein im Kosmos? *Neue Wissenschaft* 15 (1/2): 1–17.

Hoerner, Sebastian von. 2003. *Sind wir allein? SETI und das Leben im All*. München: C. H. Beck.

Hurst, Matthias. 2004. Stimmen aus dem All – Rufe aus der Seele. In *Der maximal Fremde. Begegnungen mit dem Nichtmenschlichen und die Grenzen des Verstehens*, Hrsg. Michael Schetsche, 95–112. Würzburg: Ergon.

Hurst, Matthias. 2008. Dialektik der Aliens. Darstellungen und Interpretationen von Außerirdischen in Film und Fernsehen. In *Von Menschen und Außerirdischen. Transterrestrische Begegnungen im Spiegel der Kulturwissenschaft*, Hrsg. Michael Schetsche und Martin Engelbrecht, 31–53. Bielefeld: transcript.

Jastrow, Robert. 1997. What Are the Chances for Life? *Sky & Telescope* June 1997: 62–63.

Jones, Morris. 2013. Mainstream Media and Social Media Reactions to the Discovery of Extraterrestrial Life. In *Astrobiology, History and Society: Advances in Astrobiology and Biogeographics*, Ed. Douglas Vakoch, 313–328. Berlin: Springer.

Kuiper, T. B. H. & M. Morris. 1977. Searching for Extraterrestrial Civilizations. *Science* 196 (May 6, No. 4290): 616–621.

McConnell, Brian. 2001. *Beyond Contact. A Guide to SETI and Communicating with Alien Civilisation*. Sebastopol, CA: O'Reilly.

Michaud, Michael A. G. 1999. A Unique Moment in Human History. In *Are We Alone in the Cosmos? The Search for Alien Contact in the New Millennium*, Eds. Byron Preiss & Ben Bova, 265–284. New York: ibooks.

Michaud, Michael A. G. 2007. *Contact with Alien Civilizations. Our Hopes and Fears about Encountering Extraterrestrials*. New York: Springer.

Müller, Klaus E. 2003. Tod und Auferstehung. Heilserwartungsbewegungen in traditionellen Gesellschaften. In *Historische Wendeprozesse. Ideen, die Geschichte machten*, Hrsg. Klaus E. Müller, 256–287. Freiburg: Herder.

Müller, Klaus E. 2004. Einfälle aus einer anderen Welt. In *Der maximal Fremde. Begegnungen mit dem Nichtmenschlichen und die Grenzen des Verstehens*, Hrsg. Michael Schetsche, 191–204. Würzburg: Ergon.

Panovkin, Boris Nikolaevich. 1976. The Objectivity of Knowledge and the Problem of the Exchange of Coherent Information with Extraterrestrial Civilizations. In *Philosophical Problems of 20th Century Astronomy*. Moscow: Russian Academy of Sciences: 240–265

Peters, Ted. 2013. Would the Discovery of ETI Provoke a Religious Crisis? In *Astrobiology, History and Society: Advances in Astrobiology and Biogeographics*, Ed. Douglas Vakoch, 341–355. Berlin: Springer.

Pooley, Jefferson D. 2013. Checking Up on the Invasion from Mars: Hadley Cantril, Paul Felix Lazarsfeld, and the Making of a Misremembered Classic. *International Journal of Communication* 7: 1920–1948.

Rausch, Renate. 1992. Der Kulturschock der Indios. In *1492 und die Folgen: Beiträge zur interdisziplinären Ringvorlesung an der Philipps-Universität Marburg*, Hrsg. Hans-Jürgen Prien, 18–32. Münster, Hamburg: LIT.

Schetsche, Michael. 2005. Zur Prognostizierbarkeit der Folgen außergewöhnlicher Ereignisse. In *Gegenwärtige Zukünfte. Interpretative Beiträge zur sozialwissenschaftlichen Diagnose und Prognose*, Hrsg. Ronald Hitzler und Michaela Pfadenhauer, 55–71. Wiesbaden: VS-Verlag für Sozialwissenschaften.

Schetsche, Michael. 2008. Auge in Auge mit dem maximal Fremden? Kontaktszenarien aus soziologischer Sicht. In *Von Menschen und Außerirdischen. Transterrestrische Begegnungen im Spiegel der Kulturwissenschaft*, Hrsg. Michael Schetsche und Martin Engelbrecht, 227–253. Bielefeld: transcript.

Schetsche, Michael. 2012. Theorie der Kryptodoxie. Erkundungen in den Schattenzonen der Wissensordnung. *Soziale Welt* 63 (1): 5–25.

Schetsche, Michael, René Gründer, Gerhard Mayer & Ina Schmied-Knittel. 2009. Der maximal Fremde. Überlegungen zu einer transhumanen Handlungstheorie. *Berliner Journal für Soziologie* 19 (3): 469–491.

Schetsche, Michael & Ina Schmied-Knittel. 2018a. Zur Einleitung: Heterodoxien in der *Moderne*. In *Heterodoxie. Konzepte, Traditionen, Figuren der Abweichung*, Hrsg. Michael Schetsche und Ina Schmied-Knittel, S. 9–33. Köln: Herbert von Halem.

Schetsche, Michael & Ina Schmied-Knittel. (Hrsg.) 2018b. *Heterodoxie. Konzepte, Traditionen, Figuren der Abweichung*. Köln: Herbert von Halem.

Schmied-Knittel, Ina & Michael Schetsche. 2003. Psi-Report Deutschland. Eine repräsentative Bevölkerungsumfrage zu außergewöhnlichen Erfahrungen. In *Alltägliche Wunder. Erfahrungen mit dem Übersinnlichen – wissenschaftliche Befunde*, Hrsg. Eberhard Bauer und Michael Schetsche, 13–38. Würzburg: Ergon.

Schmitz, Michael. 1997. *Kommunikation und Außerirdisches. Überlegungen zur wissenschaftlichen Frage nach Verständigung mit außerirdischer Intelligenz.* (Unveröffentlichte Magisterarbeit, Universität-Gesamthochschule Essen).

Schrogl, Kai-Uwe. 2008. Weltraumpolitik, Weltraumrecht und Außerirdische(s). In *Von Menschen und Außerirdischen. Transterrestrische Begegnungen im Spiegel der Kulturwissenschaft*, Hrsg. Michael Schetsche und Martin Engelbrecht, 255–266. Bielefeld: transcript.

Sheridan, Mark A. 2009. *SETI's Scope: How the Search for Extraterrestrial Intelligence Became Disconnected from New Ideas about Extraterrestrials.* Ann Arbor: ProQuest.

Shostak, Seth. 1998. *Sharing the Universe. Perspectives on Extraterrestrial Life.* Berkeley: Berkeley Hills Books.

Shostak, Seth. 1999. *Nachbarn im All. Auf der Suche nach Leben im Kosmos.* München: Herbig.

Stagl, Justin. 1997. Grade der Fremdheit. In *Furcht und Faszination – Facetten der Fremdheit*, Hrsg. Herfried Münkler, 85–114. Berlin: Akademie Verlag.

Stenger, Horst. 1998. Soziale und kulturelle Fremdheit. Zur Differenzierung von Fremdheitserfahrungen am Beispiel ostdeutscher Wissenschaftler. *Zeitschrift für Soziologie* 27 (1): 18–38.

Strugazki, Arkade & Boris Strugazki. (russ. Orig. 1971) 1975. *Picknick am Wegesrand. Utopische Erzählung.* Berlin: Verlag Das neue Berlin.

Sukhotin, Boris Viktorovich. 1971. Methods of Message Decoding. In *Extraterrestrial Civilizations. Problems of Interstellar Communication*, Ed. S. A. Kaplan, 133–212. Jerusalem: Keter Press.

Waldenfels, Bernhard. 1997. *Topographie des Fremden. Studien zur Phänomenologie des Fremden*, (Band 1). Frankfurt am Main: Suhrkamp.

Wendt, Alexander & Raymond Duvall. 2008. Sovereignty and the UFO. *Political Theory* 36 (4): 607–633.

The Cultural Preparation for the First Contact

9

In this chapter we go beyond what has been regarded as the task of sociology (i.e. also of its subdisciplines) in recent decades in two respects. In some of the previous chapters we had already left behind the retrospective analysis common today and the mostly tolerated diagnosis of the present in favour of prognostics understood in the classical sense of futuro-logic.[1] Here we now go one step further and, following a diagnostic description of the situation and a small review of the literature, make some suggestions as to what *could and should* happen in the near future so that the negative effects of humanity's first contact with an extraterrestrial intelligence that we have predicted cannot be avoided entirely (this seems unrealistic to us), but can at least be minimized. At this point at the latest, we leave the purely scientific level of observation—and become, as it were, cosmo-political.

9.1 General Problems and Deficits

The results of the scenario analysis we carried out give cause for concern: Situations are conceivable in which considerable *risks for mankind as a whole emanate* from the first contact—which some SETI researchers seem to be eagerly hoping for to this day. At best, that contact is an economic and ideological disruptive event; at worst, it could lead us into another major war or, in the case of the encounter scenario, under unfavorable circumstances, even to the extinction of humanity. Therefore, it is all too legitimate to ask whether and, if so, how such

---

[1] The discussions in the anthology by Hitzler and Pfadenhauer (2005) show how problematic the perception of prognostic procedures is in sociology today (in stark contrast to the self-positioning of the discipline in the 1970 and 1980s).

negative consequences can be prevented. To put it in concrete terms: Can measures already be taken today to at least reduce the negative effects of first contact? Before we go into the measures discussed in the scientific literature so far, we would like to make a few general observations from a sociological perspective, which are necessary to understand why we consider the situation to be rather precarious from a prognostic point of view, despite various detailed proposals that have been put forward in recent decades. Our central thesis here is that *humanity is anything but well prepared for first contact with an extraterrestrial intelligence.* On an abstract level, five general deficits can be named:

1. **Our thinking about extraterrestrials is shaped by anthropocentric prejudices**: We have already dealt with this in detail in Chap. 4 and therefore do not need to repeat it here. The only thing to keep in mind is that preconceptions of any kind are likely to interfere with unbiased contact. Since we are and remain human beings, we necessarily always see 'the others' with our human eyes and think about them with our human brains. The central task of exosociology in this context is to make what we believe is a perfectly possible distinction between *avoidable and unavoidable anthropocentrisms.* Just as we in the social and cultural sciences are currently trying to reflexively come to grips with our Eurocentric tradition of thought,[2] we must overcome our anthropocentrism in order to be in a position to recognize that 'terracentrism' at all which leads us to believe that life on other planets must be similar to that on Earth and must also have developed in exactly the same way. We predict now that we are in for one or two big surprises with regard to the strangeness of an extraterrestrial intelligence. To reduce the likelihood of 'nasty surprises', one of the things we should do is to change the way we think about aliens and alienness in general. (At this point, the position of exosociology as the— to date rather illegitimate—child of strangeness research also becomes clear once again).

2. **We lack an advanced theory of foreign understanding and communication with non-human actors**: We would already have the possibility to train our foreign understanding on the basis of communication with other very complex species on our planet (such as primates, dolphins or crows)—and in the near future, dealing with artificial intelligences created by us will probably also

---

[2] We are thinking here in particular of the debates conducted internationally under the heading of 'postcolonialism' (for an overview, see Kerner 2012 and Castro Varela and Dhawan 2015).

be helpful here. A central building block for this is likely to be the development of a general—and in the cultural studies sense reflexive—theory of *foreign understanding and interspecies communication*[3], a theory that can be empirically tested and developed step by step. Such a theory, probably rather a whole complex of theories, should not only benefit us in the case of communication with extraterrestrial intelligences, but will even before that facilitate our everyday interaction with wild, farm and domestic animals as well as with the increasingly common robots and autonomous machines. However, a prerequisite for the formulation of an elaborate complex of theories on foreign understanding and interspecies communication is that we set up appropriate, strongly interdisciplinary research programmes involving representatives of linguistics and behavioural biology, anthropology and sociology, psychology and cognitive science, computer science and robotics, as well as other scientific professions. Such programs also include the establishment of appropriate scientific communication outlets—such as relevant journals. Our impression is that, in view of all the (supposedly or actually) 'purely human' problems on our planet, the willingness to deal with the non-human other tends to decrease in the sciences, but also in society at large.[4] This is likely to take its revenge at the latest in the case of first contact.

3. **The problem of first contact has so far been ignored by policy-makers in international bodies**: The political agenda-setting of any era depends on a whole range of factors (Hilgartner and Bosk 1988; Brosius 1994). For our very specific context, three factors seem to be of particular importance: (1) The orientation of political elites to parliamentary cycles, in Germany usually four years long, makes it difficult to put issues on the agenda that point significantly further into the future. The political elites in post-war Germany have only been interested in futurological issues in the broadest sense for a very short window of time: With the beginning of the social-liberal coalition under chancellor Willy Brand, we find for a few years the tendency to systematically scientify state planning by resorting to futurological methods. Not least under the influence of the volume *The Limits to Growth* (Meadows 1972), a political orientation towards the future emerged, which seriously asked about the long-term transformation of society and, not least, about the consequences

---

[3] The "theory of the maximum stranger" proposed by one of us years ago (Schetsche 2004) can only be a first small building block for such a theory building.

[4] Our hope is that in the coming years and decades, AI research and the use of more and more autonomous machines will change this situation from the ground up.

of catastrophic misdevelopment. This perspective of government action, however, quickly degenerated into legitimizing rhetoric. And where the orientation of political action, as in the case of the climate problem, is indispensable to a distant future, the short-term interests of states and multinational corporations have made and continue to make rational decisions virtually impossible. In one sentence: Future and even hypothetical problems have virtually no chance of being placed on the political agenda because of the way political decision-making processes are organised in contemporary societies. (2) Due to the public ridicule[5] of all questions that have to do with 'extraterrestrials' in the broadest sense (Jüdt 2013), it is extraordinarily difficult for the political-administrative system, not only in Germany, to put corresponding questions on the political agenda by itself. Viewed correctly, we must state here a process of *negative agenda-setting*, in which in particular the leading media, which are significant for the formation of public opinion, prevent a serious discussion about the consequences of first contact with extraterrestrials from developing that can have an effect in the political sphere. Any federal government or even any EU institution that would spend public money on such research projects would have to reckon with persistent and sharp public criticism.[6] Experience shows that political decision-makers are only exposed to such criticism if the issue affects their own very specific interests (or those of their region of origin). Neither the one nor the other is the case here. (3) There is no lobby organization for dealing with the topic 'extraterrestrials/first contact'. The SETI researchers in the USA had to experience this already in 1993, when the governmental support of their projects was stopped due to the political pressure of a few conservative congressmen. Apart from the researchers themselves, there were virtually no organisations and only a few individuals who would have supported the research in question. Even their own profession, radio astronomy, watched the discontinuation of the corresponding projects with disinterest to competitive benevolence (cf. Garber 1999; Michaud 2007, pp. 39–40). Since then, the corresponding programs in the USA have been exclusively privately funded. And in the EU, as far as we can survey the research landscape, there has been no significant research at all

---

[5] For general comments on the public ridicule of subjects perceived as deviant (heterodox), see Schetsche (2013).

[6] Ultimately, probably a psychodynamic process of collective defense against extremely frightening thoughts—but we cannot pursue this idea further at this point.

on the question of first contact with extraterrestrial civilizations to date.[7] Even the international rise of the new discipline of astrobiology was largely slept through in Europe—we are years, rather decades, away from the realisation of SETI or even SETA projects.

4. **We have no binding international rules for dealing with extraterrestrials**: The consequence of the refusal of state and semi-state institutions to deal even marginally with the hypothetical problem of first contact is the absence of any binding *legal norms that* could regulate and guide state and especially international action in this field. So far, there is only one agreement of the responsible scientific society for the field of SETI research: The "Declaration of Principles Concerning the Conduct of the Search for Extraterrestrial Intelligence" was adopted in 1989 by the responsible working group of the "International Academy of Astronautics". It is not a legally binding contract, but rather a kind of self-commitment of the researchers working in this field.[8] This so-called *Post Detection Protocol* is based on the SETI paradigm valid today and primarily regulates the handling of a (potential) signal of an extraterrestrial intelligence that was received by radio telescopes or similar technical facilities. Only one sentence deals very abstractly with the question of a possible answer: It may not be broadcast without the approval of a representative international organization such as the UN.[9] On the other hand, the declaration does not contain any regulations for METI projects, i.e. for sending radio messages or similar signals directed at possible extraterrestrial civilizations somewhere in the universe. In our opinion, however, the main problem lies not in the incompleteness and far too abstract formulation of the relevant provisions, but in the fact that it is a *voluntary* agreement to which ultimately no one is bound. The corresponding regulation is guaranteed neither by national nor

---

[7] "If you look at the current space policies of Germany and France or, more generally, at the 'European Space Policy', you will search in vain for the topic of 'extraterrestrial intelligence' there" (Schrogl 2008, p. 255). Since then, it should be added, this observation has not changed. The question of extraterrestrial intelligence is, as it were, a kind of 'no-go-area' of European research policy.

[8] An updated version was adopted in October 2010 (see Baxter and Elliot 2012; Bohlmann and Bürger 2018); the current version of the Declaration can be found at: http://avsport. org/IAA/protocols_rev2010.pdf (Accessed: 10 October 2017). There was a similar self-commitment by US and Soviet researchers as early as 1971, but it has since been largely forgotten (see Harrison 1997, pp. 256–257).

[9] "Response to signals: In the case of the confirmed detection of a signal, signatories to this declaration will not respond without first seeking guidance and consent of a broadly representative international body, such as the United Nations" (online source, see footnote 8).

by international law, and there is no threat of any consequences in the event
of a breach (Schrogl 2008, p. 255). This non-regulation has significant consequences for all the basic scenarios of first contact that we examined in the
previous chapter: In the case of a signal contact, it is completely unclear who
makes a binding decision about a possible response (and who, if necessary,
stops private initiatives that believe they can speak 'for humanity'). When an
artifact is found, it is unclear who gets the power of disposal over it, who is
allowed to examine it, and who—perhaps the crucial point—decides whether
the object(s) will be brought to Earth. And in the case of a direct encounter,
perplexity regarding the question of who should attempt to communicate with
the alien intelligences is, as it were, pre-programmed. Thus, even today, one
must agree with the conclusion of Schrogl (2008, p. 261), which is now ten
years old:

> If aliens were to visit Earth today, or at least send a message, we would not be
> very well prepared and the governmental and international entities would probably
> behave as we (and they) have seen in recent American motion pictures. Such a
> scenario would probably not end well [...].

In view of the explosive nature of first contact, which we have discussed in
detail in various chapters, this also appears to us to be the central deficit: there
is a lack of binding international rules regarding jurisdiction for this 'case of
the cases'.[10]

---

[10] For the development of a "SETA Post Detection Protocol" the science author Tobias D.
Gerritzen has made first preliminary considerations: "The discovery of an extraterrestrial artifact on planets or moons of our solar system or in the Earth-Moon system could result in
extremely negative reactions of the terrestrial society. The reason would be that humanity
would suddenly have to realize that it may have been discovered a long time ago by a very
advanced alien species. This uncertainty about human reactions to physical contact by an
alien artifact makes a SETA protocol to manage the discovery and information dissemination
process long overdue. [...] The protocol is intended to provide guidelines to government and
private space agencies or corporations, as well as policy makers, on how to handle the discovery of an extraterrestrial artifact (astroengineering, passive artifacts, self-replicating artifacts,
active robotic probes). In doing so, the protocol is not intended to be applicable only to future
targeted programs to search for extraterrestrial artifacts. Rather, it is also intended to cover
potential discoveries by manned or unmanned space missions, as well as orbital monitoring
of space debris or Near Earth Objects" (T. D. Gerritzen. Personal communication: e-mail to
M. Schetsche dated 20–31-2017).

5. **We do not have a prepared contact zone at a greater distance from Earth**: we have already made it clear in various sections of the previous chapter that there are some parameters to be called critical in both the artifact and encounter scenarios. This concerns both artifacts with a recognizable or at least presumed functionality and direct encounters with intelligent entities (be they of biological or other quality). In both cases, Earth itself seems to us a (too) risky place for investigation and encounter. This statement comes without any presuppositions regarding assumed 'evil intentions' of extraterrestrials, as we know them sufficiently from science fiction of the last decades. An artifact found in space, if brought to Earth, can have effects that are highly negative for life on Earth (human and non-human), completely independent of the motives and specifications of its creators (Baum et al. 2011, pp. 2124–2126; Baxter and Elliott 2012, p. 34). On the one hand, this concerns potential contamination with microorganisms that could negatively affect the terrestrial biosphere,[11] but on the other hand, it also concerns possible modes of operation of the artifact that are latent and could be activated during an investigation. If we assume that such artifacts are produced by a civilization that is far advanced compared to humanity, even comparatively small artifacts could have significant negative effects on the environment—already in the intended normal operation, even more so if malfunctions are triggered by improper, namely human, investigations. To give an almost trivial example: An exotic drive intended for free space, activated on a planetary surface, could cause massive damage to the environment—an environment that could well encompass an entire continent. All of this applies to an even greater extent to the visit of an intelligence-controlled spacecraft to Earth—both independently and depending on the motives of the intelligence in question.[12] The spectrum ranges from unknowing contamination of the

---

[11] Today, this problem is primarily discussed in the context of the "Planetary Protection Protocols", which have been adopted by various space organisations. On the one hand, this concerns the protection of possible extraterrestrial life from contamination with terrestrial microbes in the context of space probe missions, but also the protection of the terrestrial biosphere from extraterrestrial organisms that could be introduced, as it were, during missions that have collected samples from alien celestial bodies (Rummel 2001; Spry 2009). To our knowledge, however, the problem of dealing with extraterrestrial artifacts has been regularly overlooked in the process; the relevant regulations and procedures should urgently be expanded.

[12] "The most specific nature, of a security threat would be the approach of alien space vehicles to our solar system without acceptable guarantees of non-hostile intentions. If we did not have the capability to intercept or neutralize these vehicles at great distance and did not know the type or range of weapons they might carry, we would have to try placing the solar system

environment to unintended effects of the actions of extraterrestrials on the terrestrial environment to intentional destruction (if the first contact is not quite as peaceful and friendly as the contact optimists hope for). For these reasons, it seems indispensable to us to establish strict rules for dealing with extraterrestrial artifacts and space probes: Under no circumstances should extraterrestrial artifacts found in space be brought to Earth for study, or even to near-Earth space. And active spacecraft[13] should be offered a hard-to-miss spatial contact, as far outside Earth's orbit as possible. A place for such contacts or even for the investigation of unknown artifacts does not yet exist—and as far as we know the plans of the space nations and space companies for the next decades, such considerations do not play a role there so far. This seems to us a serious deficiency—we will return to this question later.

9.2   Some Suggestions from the Scientific Debate

Because of these problems and deficits, we believe that a discussion of possible and necessary measures to prepare for first contact is urgently needed. To date, however, the scientific literature on the subject of extraterrestrial intelligence has only rarely addressed this complex set of questions. The 1960 *Brookings Report* commissioned by NASA (which was obviously far ahead of its time) had already raised the question of what measures could be taken to prepare for first contact—especially with regard to a planned education of the public. In particular, he had suggested that appropriate scientific studies be conducted (Brookings Report 1960, pp. 182–184[14]). To our knowledge, however, detailed discussions on concrete measures did not take place in the English-speaking world until 1991 and

---

off limits by negotiation, perhaps setting up a no man's land between the stars until acceptable rules of visitation were worked out" (Michaud 1972, p. 14). Cf. also the considerations in Korhonen (2012).

[13] Whether these rules can be communicated and enforced in contact with an intelligence-controlled spacecraft is another question. Humanity is unlikely to be able to prevent the spacecraft of a technologically advanced civilization from entering Earth's atmosphere or even landing on Earth—but this becomes more likely the fewer off-planet contact alternatives there are. We *speculate* here once in the direction that from the point of view of extraterrestrial visitors the lack of compatibility between different planetary atmospheres as well as the chemical, biological and also social risks, which possibly cannot be completely assessed by the strangers, speak against a meeting directly on the earth, provided that other contact possibilities at easily attainable places in the solar system are recognizable.

[14] Page numbers in original (in the transcript used: pp. 215–217).

1992—in the context of a three-part workshop organized by the SETI Institute under the heading "Social Implications of the Detection of Extraterrestrial Civilization". The results of the discussions at that time are documented in a small conference volume (Billingham et al. 1994). In accordance with the then (and now) paradigmatic orientation of the SETI Institute, the discussions assumed that first contact would take the form of what is, according to our analytical scheme, a 'distant remote contact'. The participants of the conference suspected that already such a signal reception would be sufficient to lead to serious cultural and especially religious upheavals on Earth. Accordingly, they proposed a whole series of 'cushioning' preparatory measures that could and should be initiated even before the event.

The starting point of all proposed measures (Billingham et al. 1994, p. 80–81) was the consideration that the very different *cultural backgrounds* would play a decisive role in how the population would deal with the information about the first contact: There would be cultures in which the idea of the existence of extraterrestrials is widespread, while in other cultures their existence is almost unthinkable. As a *scientific preparation strategy*, the discussion participants therefore demanded, among other things, the implementation of appropriate comparative cultural studies as well as the identification of subgroups with differing alien images and first contact ideas. In addition, better use should be made of all types of mass media to inform the public about the possibility of the existence of extraterrestrials and SETI research. The implicit basic thesis here was that the public's reaction to first contact would depend on their level of 'enlightenment': the more people knew about SETI projects and the possibility of intelligent extraterrestrial life, the more level-headed and, as it were, 'composed' their reaction would be if the worst came to the worst. It did *not* occur to the participants that this *could* also be the case the other way round—at least such fears are not documented in the present conference proceedings.

Billingham et al. (1994, p. 94) specifically proposed *six measures* to prepare humanity for the reception of an extraterrestrial signal: (1) establishing a "post-detection protocol",[15] (2) establishing procedures by which policy makers and the public can be informed of first contact made, (3) keeping governmental and international institutions informed of the status of SETI research, (4) conducting systematic analyses of the policy implications of first contact, (5) developing national and international procedures governing how a signal might be responded to, and (6) expanding international participation in SETI projects.

---

[15] This measure had already been implemented at the time of the three workshop sessions, at least in terms of tendency—we had already referred to this in Sect. 9.1.

In the section "SETI, Education, News, and Entertainment", the participants also made very far-reaching suggestions for 'educational' measures which should prepare mankind for a signal contact and also inform it in the event of an incident:

1. Schools, universities, museums, libraries[16] and other educational institutions should inform especially the young generation about the possibilities of extraterrestrial life and the chance of first contact.
2. The mass media should systematically report on SETI programs so that the media themselves (i.e. their reporters, commentators, etc.), but also their recipients, are prepared when first contact occurs.
3. The entertainment media, but also artists of any kind, should inform about the possibility of a (peaceful) first contact through corresponding projects and thereby promote SETI research; after the contact has taken place, they should inform the population about the extraterrestrials as well as the communication attempts in an entertaining way.

For each of these three areas, a series of sometimes more, sometimes less concrete proposals were made—and the respective possibilities of the most diverse institutions, media and individuals to make a contribution to enlightenment were evaluated (Billingham et al. 1994, pp. 95–119). In doing so, and this cannot be stressed often enough, all proposed measures were strictly oriented towards—from our point of view least consequential—distant contact. Measures for other contact scenarios were not even considered, probably simply because these scenarios were downright *unthinkable* according to the dominant SETI paradigm at the time.

In the following years (today one must almost write: decades) only a few scientific works have appeared that have dedicated themselves to the question of the concrete preparation for a first contact with even a rudimentary degree of

---

[16] Mention should also be made here of the criticism (Billingham et al. 1994, p. 125) that in many libraries there is a disproportion between 'serious' SETI literature on the one hand and 'pseudoscientific' literature, for example on UFO issues, on the other. The aim here is obviously to clean up public discourse so that information can only be provided about the possibility of distant contact, while other contact scenarios (to which the ET hypothesis in UFO research ultimately *also* refers) are to be excluded as far as possible. From the perspective of the history of science, this proves once again (Romesberg 1992, passim) that many SETI activists pursue a 'hidden agenda' aimed at excluding all knowledge about extraterrestrials and first contact that does not follow the SETI paradigm they represent. For decades they have been trying to establish a kind of *first contact orthodoxy*.

detail.[17] One of the few exceptions is the internationally acclaimed work *Contact with Alien Civilisation* by the long-time professional diplomat (Michaud 2007). In his book there is an Appendix in which the author offers twenty pages of suggestions on how to prepare for a possible first contact. His suggestions, sometimes very basic, sometimes quite concrete, can be summarized in seven thematic or question complexes:

1. The need to adopt *internationally binding rules* on what should happen in the event of a first contact (the author here calls for agreements for all three types of baseline scenarios we have studied: Signal, Artifact and Encounter scenarios[18]).

2. The development of a *formal scale* with the help of which the significance and also the risk of each type of contact can be assessed according to different parameters and communicated to the public accordingly (the author refers here as an example to the so-called Rio *scale* presented in 2000 at an astronomical congress in Rio de Janeiro).

3. The clarification of the question of which institution (scientific, national or international) informs the *world public at* what time and in what detail about the initial contact that has taken place.

4. The preparation of political decision-makers, national and international organisations for a possible first contact (this includes basic information on the state of research as well as that on possible political and cultural consequences of the first contact).

5. The development of legally binding procedures governing who speaks for the Earth—in the case of remote or direct contact—what information should be exchanged, and how messages should be worded to minimize misunderstandings.

6. Preparing policy makers and international institutions for the possible establishment of a long-term cultural relationship with one or even more

---

[17] However, there are now a large number of popular publications on the subject of first contact, particularly in the English-speaking world, which address this issue—in greater or lesser detail. A systematic presentation of such proposals would go beyond the scope of this chapter. We will therefore limit ourselves in the following to a few scientific publications whose statements appear to us to be particularly significant.

[18] "Some have argued that the Declaration of Principles is focused too narrowly on the detection of an electromagnetic signal and does not apply to other scenarios of contact. Stride thought that we need protocols for both the search for extraterrestrial artifacts (SETA) and the search for extraterrestrial visitations (SETV). These documents would include strict rules for verification, confirmation, even syntax for communication" (Michaud 2007, p. 362).

extraterrestrial powers (including the formulation of basic principles for an extraterrestrial legal system from a human perspective).

7. Preparing for the defense of our planet, and in particular ensuring the survival of humanity if the aliens are hostile or direct contact—for whatever reason—is negative.[19]

Most recently, in the conference volume *The Impact of Discovery Life beyond Earth* (2015), edited by historian Steven J. Dick, there is an entire section with various contributions, all of which seem to be devoted to the question of how society could prepare for first contact. A closer look at those six papers, however, reveals that only two of them actually address this problem[20]—and one of them is again by Michael A. G. Michaud.

Since we have already discussed his ideas on contact preparation in detail in the 2007 book, we will focus here on the expositions of Race (2015): *Preparing for the Discovery of ET Life*. The author begins by noting that any planning of first contact preparation measures must assume that there are very different contact scenarios, each requiring different arrangements (Race 2015, p. 264). However, Race is referring here to the implications of the discovery of life in general, and is concerned *exclusively* with remote contact in line with the SETI paradigm when considering contact with intelligent life forms.[21] In her account, she focuses

---

[19] The author here explicitly contradicts the assessment (also shared by us) that a military conflict with an extraterrestrial power is always likely to be hopelessly asymmetrical: "SETI conventional wisdom assumes that because we will be much less technologically advanced than any other civilization the we contact, we would be helpless if the extraterrestrials were hostile. This disparity may turn out to be true, but it remains unproven. To assume our weakness in advance would be preemptive capitulation" (Michaud 2007, p. 376).

[20] It is highly interesting from a sociological point of view that these two contributions (at the same time the last of the conference proceedings) are devoted to the question of what the social consequences would be if the search for extraterrestrial intelligence were to remain unsuccessful in the long term (Billings 2015; Chaisson 2015). It is possible that a growing scepticism about the prospects of success of traditional SETI strategies is emerging here.

[21] The remarks in the contribution by Klara Anna Capova (2013) are similar—where it is not a matter of microorganisms but of intelligent extraterrestrial life, the author considers exclusively a remote contact in accordance with the SETI paradigm. On the question of cultural reactions to a first contact of this kind, she emphatically points to the influence of alien imagery from science fiction—which is definitely seen as problematic. Her conclusion: "If the ETL [Extraterrestrial Life] debate is to be moved forward, a better understanding needs to be developed of the cultural landscapes from which the reaction of the public to the detection of extraterrestrial life arises. We can speculate on the possible wider implications of our narratives about the encounter with aliens. But to be clear, we must cautions about making any generalizations independent of the specific contact situation. The immediate societal

on presenting the legal and institutional frameworks that have been in place *to date to* enable a *rational* response to the discovery of extraterrestrial life. Those regulations and plans seem to her sufficient, if too strongly formulated from a scientific point of view:

> With no know direct impacts involved and very long lag times for interacting with presumed ET life, the consequences of detection will not likely have any real-time effects upon continuing search activities. In contrast, while astrobiologists are prepared in general for interpreting scientific aspects of possible discoveries, there are a number of gaps in their plan. (Race 2015, p. 281)

However, this relativization does not lead to the consequence of presenting more detailed and concrete proposals in case of contact with extraterrestrial intelligences. Their adherence to the SETI paradigm with corresponding long periods of time that seem to be available for discussions about the possible reactions to a signal reception makes any preparation that goes beyond the previous non-binding protocols superfluous from their point of view.

We have paid attention to this contribution primarily because it makes clear the consequences of a deliberately limited scope of possibilities in first contact scenarios: Those who only consider the classical SETI scenario have little to worry about preparatory measures. As will become apparent in a moment (in the next subsection), we broadly agree with this assessment. The situation changes completely, however, if we consider alternative contact scenarios-which in that volume Michael A. G. Michaud does exclusively. We had presented his concrete proposals for preparatory measures—a very general, but in our opinion important addition is provided in the newer, much shorter text only by point 14 of the list of measures:

> **Recognize the limitations of prediction.** We humans are notoriously inaccurate when we predict the future more than a few years ahead. [...] We do not know which scenario of contact we will face. At best, our planning will be only partly successful (Michaud 2015, p. 295; emphasis in original).

Caution is also advised by Baxter and Elliott (2012, p. 33) in their political science essay *A SETI Metapolicy*. They strongly suggest that the discovery of an extraterrestrial civilization is likely to have significant *security* implications for

---

response to the detection of extraterrestrial life will be cultural as well as individual, but above all contextual, and in any case influenced by the type of life discovered" (Capova 2013, pp. 278–279).

the entire planet: Despite the vast interstellar distances, wars between civilizations are conceivable. The authors believe that this factor must be considered,
especially when deciding on METI programs. They also (this is a rare suggestion
in the literature!) explicitly suggest that critical *infrastructure on Earth* should be
designed to withstand not only natural disasters but also an attack by an extraterrestrial intelligence. The conclusion of their unusual contribution is that "The
most robust policy principle may be to hope for the best from the contact, but to
prepare for the worst" (Baxter and Elliott 2012, p. 35).

9.3   Conceivable Strategies for Minimising the Negative Effects of First Contact

If one systematizes these quite different suggestions for preparation strategies,
which all aim at minimizing possible *negative effects of the first contact*, four
basic types can be distinguished:

   I. Strategies to *reduce the likelihood* of first contact (an upfront, ultimately very
      radical approach);
  II. Strategies for *keeping secret* a first contact that has taken place[22];
 III. Strategies for influencing public opinion before and after the event becomes
      known (*crisis communication*);
  IV. Strategies to minimise harm if contact is conflictual or risky situations of
      other kinds arise (*safety precautions*).

In the following compilation, we again differentiate according to the three basic
scenarios we presented in the previous chapter. It should also be pointed out
that—in contrast to the descriptions in Billingham et al. or Michaud, for example—we are less concerned with concrete proposals than with a discussion of the
fundamental possibilities that exist for taking away some of the predicted cultural
acuteness of first contact. In order to prevent this sub-chapter from becoming too
extensive, we have to limit ourselves to key points in some places—this makes

---

[22] Here, a cosmo-political secret in the true sense of the word would emerge, with all the
problems associated with processes of secrecy (Schetsche 2008).

all the more sense here and there, as some of the technical strategies mentioned are located far outside of social science competence.

I. Reducing the likelihood of contact

Signal scenario
Two possible goals must be distinguished here. If the goal is that we ourselves do not want to receive extraterrestrial signals, the simplest measure is to immediately stop all SETI projects and to decide as a world society to no longer pursue such research (for example, because the majority considers it too risky). Even if the previous SETI projects were one hundred percent unsuccessful (at least as far as their primary goal, the discovery of extraterrestrial intelligence, is concerned[23]), it cannot be assumed that researchers all over the world would bow to such a vote. Quite apart from the fact that it is completely open who should issue such a ban. Not least against the background of the limited impact of a success of these projects, as we have predicted, it made little sense to envisage such a measure. The situation is different with the *METI projects*, which attempt to draw the attention of other civilizations to Earth or to humanity by means of signals deliberately sent out into the vastness of the cosmos. The general problem of this aim we had discussed in detail elsewhere (Chap. 6). Here it is only necessary to add that, apart from generally refraining from such "high-risk experiments", various technical measures are conceivable to reduce the probability of an *accidental* discovery of the Earth by an alien intelligence. However, we are not technically competent enough to decide whether such measures (such as the development of hard-to-detect communication channels) are promising and can be implemented at reasonable cost.[24]

---

[23] The secondary goal, on the other hand, of stimulating scientific, philosophical, and even public debates about man's place in the cosmos, was achieved.

[24] We do not want to speculate here about the development of camouflage techniques to hide civilization markers shining in the electromagnetic spectrum—but allow us to point out that this could well be an explanation for the failure of all SETI projects so far: Other civilizations do not *want to* be discovered. This idea can already be found in Bell (1973, p. 349)—it leads directly to the highly justified question of what reasons there might be for such behaviour of technologically advanced civilisations (Baum et al. 2011, p. 2116; Korhonen 2012).

Artifact Scenario

The most promising strategy here is certainly to 'carry on as before'. Since we consider the probability of an *accidental* discovery of at least small artifacts[25] to be rather low (even if we had chosen such a narrative in the scenario analysis presented by us above), it would require systematic search efforts to discover corresponding objects in the vastness of our solar system. The best strategy in this context would therefore be an international ban on all SETA projects—which at the moment would not even lead to an audible outcry in the scientific community, since almost nobody is engaged in this search anyway.[26] The question is whether scientific opposition would be greater if an artifact found by accident were immediately destroyed (say, under UN supervision). However, attempting to enforce a corresponding international regulation could well lead to the opposite of the intended procedure: The corresponding first contact scenario and the associated search strategies would attract scientific and also public attention to an unprecedented extent and thus possibly counteract corresponding efforts. Probably, therefore, an 'elimination' of the problem by clandestine action of governmental or private institutions would be more promising (see on this below the strategy of secrecy).

Encounter scenario

With the present state of human space technology, for the unforeseeable future, the others are the discoverers, we are merely the discovered. With the expected far superior state of extraterrestrial technology, we see no way to prevent our discovery by interstellar space probes or even the 'visit' of extraterrestrials to Earth. Here we can do no more than hope that 'they' will either adopt a strict non-interference policy of their own accord, or have gained sufficient experience

---

[25] Based on processes of increasing *miniaturization* on Earth, we assume that corresponding tendencies could also be present in extraterrestrial civilizations. If we are already planning simple interstellar space probes the size of a few centimetres (see Stirn 2016), it is conceivable that technically more advanced civilisations will be able to produce even highly complex probes with an immense range of functions in very small sizes. If we had to search for corresponding artifacts, we would look for objects less than a meter in size. This does not mean, of course, that extraterrestrial devices for certain purposes (such as resource extraction) could not be considerably larger.

[26] We had presented some theoretical proposals and debates on SETA in Chap. 5.

to make cultural contact with a 'backward' civilization (namely ours) proceed in a reasonably compatible manner.

II. Secrecy

Signal scenario
The already several times mentioned *Post Detection Protocol* (1989/2010) provides, after the technical-scientific verification of the reception of signals of an extraterrestrial civilization, first to inform various international organizations including the Secretary General of the United Nations. Only afterwards the public is to be informed via specialized and mass media. The required process of verifying the data would certainly take a long time; it would involve a whole range of research institutions with a large number of individuals. It is therefore questionable how realistic the process proposed in the Declaration is (Harrison 1997, p. 207; Shostak 1999, pp. 225–227). It is undisputed among SETI researchers that such a signal would be one of the most serious scientific discoveries of modern times—the 'news value' of such information is likely to be correspondingly high and thus the period of time until the first rumours reach the mass media and thus the public is correspondingly short (at least according to Shostak 1999, p. 226). Since from our point of view (cf. the scenario analysis in the previous chapter) the everyday effects of a corresponding message are likely to be rather small anyway, there seems to us to be little need to keep the fact of the signal reception itself secret. The situation could be different only if one day in the distant future it should be possible to decode the essential contents of the message—and if its content should prove to be threatening to mankind.

Artifact Scenario
In our estimation, concealing an artifact discovery would make rational sense under two conditions: A nation-state or corporation promises itself a monopoly in the use of alien technology, or the artifact found has an inherent message that is somehow threatening. In the former case, the concealment is primarily done in a particular interest (certainly open to cosmopolitan criticism), but may also serve to avoid conflict if there are competing powers that would appropriate the alien technology at almost any price. In the second case, situations are conceivable in which concealment would be socially and ethically justifiable, for example if the feared negative mass psychological effects could clearly exceed the expected gain in knowledge. In this case, it would even be conceivable (and possibly legitimate) to destroy the artifact in question promptly after its discovery and to erase all traces of its existence (Baxter and Elliot 2012, p. 34).

Encounter scenario
Whether there are realistic options here to keep the event secret from the world
public depends entirely on the parameters of the contact. In the case of the popular scenario in science fiction (most recently seen in the movie *Arrival*[27]—even
more threatening in the classic *Independence Day*[28]) of the conspicuous appearance of a multitude of huge alien spaceships over densely populated regions of
the Earth, this question does not arise from the outset. However, other parameters are conceivable, in which first only scientific or military research facilities
and then also the government of individual states are informed about the appearance of an obviously intelligently controlled missile. (This would be the case,
for example, if the contact took place very unspectacularly at a remote location
or even far from Earth in space). In such cases, political decision-makers might
well conclude that—for reasons of power politics or even mass psychological reasons—it would be better *not* to inform the public, at least for a while[29] (Harrison
1997, pp. 269–272). Both types of reasons are likely to be much more serious in
the encounter scenario than in other forms of first contact. Not to at least think
about such a possibility, however, seems naïve in political science terms—both
on the part of the political-administrative system and on the part of the public. If
the contact parameters allow it, concealment is a serious option.[30]

III. Crisis communication

Signal scenario
This is the area in which a large part of the proposals put forward by Billingham et al. (1994) fail, as measures to prepare for an assumed initial contact via
signal reception. We had presented the proposals in detail in the previous subsection, so we only want to point out again that two types of proposals have
to be distinguished here: Actions before a clear signal reception and Actions

---

[27] USA 2016, directed by Denis Villeneuve.

[28] USA 1996, directed by Roland Emmerich.

[29] Already the "Brookings Report" prepared on behalf of NASA had raised the question in
1961 under which circumstances and at which time the public should be informed best about
the fact of the occurred contact with an extraterrestrial intelligence.

[30] This leads us directly to the hypothesis, ubiquitous in the context of UFO research, that
first contact has already taken place but is being kept secret from the public by government
agencies. In contrast to most critics of this hypothesis, we are of the opinion that such secrecy,
while risky, is even possible for an extended period of time under certain circumstances
(Schetsche 2008, pp. 247–249). However, we do not have any convincing information that
such a case could have already occurred.

after a signal reception. In the first case, we seem to be primarily concerned with a wide variety of ideas for 'educating' the public, especially policy makers and religious leaders, about the state of scientific research and the opportunities (and also the risks) of first contact. Here, the most diverse proposals have been made in the literature on the subject (sometimes more, sometimes less 'pedagogically' thought through)—proposals that are almost universally based on the SETI paradigm and thus regrettably share its anthropocentric misconceptions. We are therefore unsure whether we should wholeheartedly wish these measures success. In the second case, namely if the signal has already been received, it is a question of influencing or controlling public opinion—an endeavor whose borderline case is the strategy of concealment. In any case, the problem is what information is transmitted to the political-administrative system at what time, and then who specifically decides whether the general public is fully (and truthfully) informed. We had already pointed out several times in the previous chapter that all attempts at precise planning of these measures seem to us to be unpromising. If the signal reception is not exactly by military agencies, the idea of being able to withhold appropriate information for more than a few days and process it appropriately first seems illusory. (We had also assumed this in the corresponding scenario analysis.) For these practical information reasons alone, it does not seem sensible to us to devote too much attention to the discussion of the corresponding measures.

Artifact Scenario
Everything just mentioned seems to us to apply—mutatis mutandis—also to the artefact scenario.[31] Special considerations would only have to be made here *after a discovery*, if the artefact(s) were either found on Earth itself or if an economic actor succeeded in gaining control over the foreign object. What the information policy might look like in the first special case depends very much on where the object is found. It is possible that its exposed location makes any debate about a systematic information policy superfluous, because a mass media 'hype' immediately arises that can no longer be controlled politically. If, on the other hand, in the second special case, a company with the power to act knows how to appropriate the object, this is likely to lead to a sustained information policy similar to that in dealing with patents, new developments, market launches etc. Here, scientific debates as well as public exchanges of opinion are likely to depend almost

---

[31] The fact that for SETA research, which deals with extraterrestrial artifacts, no strategies for 'informing the public' have yet been developed, as is usual in the SETI field, need not concern us here: The strengthening of this research perspective in the international framework, which we expect in the near future, will certainly be followed soon by the corresponding proposals for public relations.

entirely on the goodwill of the corporation in question. We cannot predict what their concrete information strategies would look like.

Encounter scenario

According to the results of the scenario analysis we conducted, this is likely to be the most culturally critical point. In the previous chapter, we had repeatedly pointed out that not only would 'close proximity contact' have by far the most lasting impact—we had also predicted that the probability of very negative effects, for example in the form of a general culture shock, would be highest here. Here, therefore, the question arises with great emphasis not only of possible 'shock-reducing' preparation strategies, but also of optimal crisis communication once the worst-case scenario has occurred.

At the beginning of any debate on *preparedness strategies,* the ethical (and also political) question inevitably arises as to whether informing the public about possible direct contact should be (a) as truthful as possible, or rather (b) pedagogically preventive. Warning indications (such as those provided not least by our own scenario analysis) could, if received by a broad public, take on the character of a self-fulfilling prophecy: The fact that 'science' warns of possible mass psychological distortions worries the public so much that those distortions come to pass in the case of first contact. The same is true for policy makers: the fantasy that the extraterrestrials come with the 'evil intention' of enslaving or destroying humanity could influence the courses of action of human actors in such a way that a conflictual course of the first encounter becomes inevitable.[32] In this case, as cautionary, rather contact-critical scientists, we can only take comfort in the fact that science fiction films like *Independence Day* are likely to influence the expectations of many people (and arguably political decision-makers) to a much greater extent than scientific forecasts ever could. Nevertheless, this does not absolve us as 'forecasters' from deciding whether we should 'prettify' our results to avoid corresponding feedbacks. We have decided against this in this volume because we think that the scientific as well as the general public should know realistic forecasts of first contact, i.e. forecasts that do not conceal conceivable negative effects. Only then can it be decided in a rational discourse whether we as human beings (here even as mankind) are prepared to accept certain risks—for example with regard to the realization of certain SETI, CETI and SETA projects.

---

[32] Causes of conflict, which are founded in the interests or aims of the extraterrestrials themselves, we leave out of consideration here for reasons of principle (but see on this already the perceptive considerations in Michaud 1972).

If direct contact has indeed become a reality, the hour of *crisis communication* has come. We cannot discuss its principles, strategic options and socio-ethical problems here (an overview of the current state of the debates is provided by Nolting and Thießen 2008; Höbel and Hofmann 2014), but merely point out that three *special features* must be taken into account in the case of interest here: (1) It is always a *global problem* in the truest sense of the word—whoever acts concretely risks worldwide consequences of action accordingly (in the globally networked media landscape, this also applies to communiqués of actors known to have the power to act). (2) What threatens at worst in the event of a failure of communication with the extraterrestrials or in the event of an unfortunate course of interaction is nothing less than the destruction of the earth and the annihilation of mankind; a greater overall risk is simply inconceivable. (3) In contrast to all other crises, humanity is confronted with an *intelligent* and *technologically superior* actor in the first contact, about whom almost nothing is known in the first place—strategies for action can therefore not rely on proven assumptions and basic rules, as they apply to interactions, for example, between human governments. Nevertheless, negative mass psychological phenomena that arise independently of (or possibly also dependent on) human interaction with the maximum stranger can be dealt with in terms of security policy and security practice according to the 'rules' for serious wild card events with other causes (such as natural disasters or terrorist attacks). The conclusions to be drawn from the special features of the first-contact situation would have to be examined in greater detail in corresponding research projects on civil protection and disaster management. Whether, beyond this, systematic preparation for communication with the maximum stranger is possible seems doubtful to us, not least because of the completely unpredictable motives, action strategies and behaviour patterns of an extraterrestrial intelligence.

IV. Safety precautions

Signal scenario
At first glance, it is neither possible nor sensible to take any measures to protect the transmitted message from negative influences when receiving a signal. This seems to be all the more true since, as has been shown several times, we consider that the chance of decoding a received message is exceedingly small. However, a second look reveals a potentially serious problem: the 'infection' of Earth's communicative infrastructure with malicious program code contained in the extraterrestrial message (Carrigan 2006; Baum et al. 2011, pp. 2125–2126; Neal 2014, p. 74; Gerritzen 2016, pp. 215–219). We are not qualified to decide

whether it is even technically possible for an extraterrestrial program to be implemented on terrestrial computers and networks—but we are also not sure that this can be completely ruled out against the background of a far more advanced computer technology of the extraterrestrials. Ultimately, it is even conceivable that a transferred extensive program code could be the extraterrestrial intelligence itself, the offshoot or clone of a comprehensive AI that forms the core of a post-biological machine civilization. As long as such a possibility cannot be ruled out, the data processing of all receiving stations should, as a matter of principle, take place on separate computers which have no connection whatsoever to computer networks, especially not to the Internet. Here, inverse precautionary measures must take effect, as they do when securing critical infrastructure against computer viruses and hacker attacks: No data packet leaves the receiving station. (However, we are not in a position to predict whether the protective measures known to us humans would be sufficient against a highly advanced foreign technology).

Artifact scenario and encounter scenario
It seems reasonable to us to deal with these two points together. As we had already stated at the beginning of this chapter (9.1, point 5), it seems too risky to us to bring extraterrestrial artifacts or visitors to Earth itself. If we have a choice (which is far from certain, at least in the second case), any investigation or communication should take place as far away from Earth as possible. And science fiction fans have suspected, at least since the movie *Life*,[33] that even Earth orbit is not a suitable place for potentially risky encounters with alien entities. The best safety precaution imaginable at the present state of our technology would therefore be the construction of an investigation and contact station far outside the Earth, for example on a Mars moon that is already easily accessible to us by space technology. Later, it could be relocated to the outer solar system to create a kind of 'cordon sanitaire' between this station and the human-populated areas of the solar system. On the other hand, if an artifact is found on Earth itself, or if an alien space probe is not attracted by offers of contact far from Earth, strict measures would have to be taken to isolate the site where it was found or landed. How these could look like, we leave to experts for 'Planetary Protection' and civil protection. And in case of doubt (here many science fiction movies are not so unrealistic according to our estimation) 'the military' will take care of these questions anyway. Whether all this will be sufficient to avert serious negative consequences of a biological, chemical or technical nature, we are not in a position to say. In our opinion, such and similar measures should be thought

---

[33] USA 2017, directed by Daniél Espinosa.

about very systematically and creatively in the near future. This is not only a task for the sciences, but also presupposes corresponding political decision-making processes in national and, in particular, international bodies.

9.4 Conclusion: Guiding Principles for Preparing for the Initial Contact

In disaster research, the magnitude of the risk of an event (such as an earthquake) is determined by two factors: its probability of occurrence and the magnitude of its negative consequences. If contact with an extraterrestrial intelligence would not only be one of the most drastic events in human history, but could also—at least under certain conditions—have devastating cultural and social consequences, the probability of the first-contact event can become almost arbitrarily low without the *overall risk* becoming negligible. For when the negative consequences of events approach infinity, only they, and not the probability of occurrence,[34] determine the risk relevance of such events (called *'low-probability, high-impact events'* in international futures research). Not to plan for such events would be irresponsible (Hiroki 2012). The ignorance displayed today by parts of the scientific community, by the public and by national and international political institutions with regard to this question only functions as an option for action because and as long as there is no obvious evidence for the existence of extraterrestrial intelligence. This strategy, however, becomes suddenly precarious at the moment when indications of intelligent life outside of the Earth accumulate or even when the first-contact event becomes unmistakable. On the basis of our scenario-analytical forecasts presented in the previous chapter, we therefore strongly recommend cultural preparation for first contact. In our opinion, this preparation should be based *on five guiding principles*:

1. The search for extraterrestrial intelligence is culturally considered high-risk research, the benefits and risks of which must be openly discussed.

---

[34] At least as long as this is not equal to zero—which hardly anyone in astro-science seriously dares to claim today.

2. Since this is a (global) risk for society as a whole, this debate must not be left to the scientific community—especially not to disciplines that associate particular interests with the topic.[35]

3. The public and political elites must be informed about this research and its possible consequences at least to the extent that rational decisions about legal regulations and the drawing of boundaries (e.g. with regard to METI projects) are possible.

4. Since *global* impacts can be assumed for all conceivable first contact scenarios, the problem primarily falls within the competence of international institutions; legal regulations and political measures should preferably be implemented at UN level.

5. In order to minimise negative effects, first contact (in *all* its probable variants) should become the subject of security research and be considered in civil defence and disaster control plans as an exceptional incident event.[36]

References

Baum, Seth D., Jacob D. Haqq-Misra & Shawn D. Domagal-Goldman. 2011. Would Contact with Extraterrestrials Benefit or Harm Humanity? A Scenario Analysis. *Acta Astronautica* 68: 2114–2129

Baxter, Stephan & John Elliott. 2012. A SETI Metapolicy. New Directions towards Comprehensive Policies Concerning the Detection of Extraterrestrial Intelligence. *Acta Astronautica* 78: 31–36.

Bell, John H. 1973. The Zoo Hypothesis. *Icarus* 19: 347–349.

Billingham, John, et al. 1994. *Social Implications of the Detection of an Extraterrestrial Civilization.* Mountain View, CA. SETI Institute.

Billings, Linda. 2015. The Allure of Alien Life. Public and Media Framings of Extraterrestrial Life. In *The Impact of Discovery Life beyond Earth,* Ed. Steven J. Dick, 308–323. Cambridge: University Press.

Brookings-Report. 1960. Proposed Studies on the Implications of Peaceful Space Activities for Human Affairs [A Report Prepared for the Committee on Long-Range Studies of the

---

[35] In this context, we explicitly mention the SETI community of radio astronomers, which, in the – also financial—interest of its own field of research, had managed for decades to largely prevent scientific and public debates about alternative, possibly even more probable contact scenarios – especially through its aggressive public relations.

[36] In this respect, the *Fire Officer's Guide to Disaster Control* (Kramer and Bahme 1992), which originates from the USA, can be regarded as exemplary. In a separate chapter, it provides instructions for the behaviour of emergency forces after an alien landing on Earth.

National Aeronautics and Space Administration by The Brookings Institution]. Washington D.C: Brookings Institution. http://www.nicap.org/papers/BrookingsCompleteRpt. pdf.

Brosius, Hans-Bernd. 1994: Agenda-Setting nach einem Vierteljahrhundert Forschung: Methodischer und theoretischer Stillstand? *Publizistik* 39: 269–288.

Capova, Klara Anna. 2013. The Detection of Extraterrestrial Life: Are We Ready? In *Astrobiology, History, and Socienty. Life Beyond Earth and the Impact of Discovery*, Ed. Douglas A. Vakoch, 271–281. Heidelberg: Springer

Carrigan, Richard A. Jr. 2006. Do Potential SETI Signals Need to be Decontaminated? *Acta Astronautica* 58 (2): 112–117.

Chaisson, Eric J. 2015. Internalizing Null Extraterrestrial ‚Signals'. In *The Impact of Discovery Life beyond Earth*, Ed. Steven J. Dick, 324–337. Cambridge: University Press.

Dick, Steven J. (Ed.) 2015. *The Impact of Discovery Life beyond Earth*. Cambridge: University Press.

do Mar Castro Varala, María & Nikita Dhawan. (2. überarb. Aufl.) 2015. *Postkoloniale Theorie. Eine kritische Einführung.* Bielefeld: transcript.

Garber, Stephen J. 1999. Searching for Good Science: The Cancellation of NASA's SETI Program. *Journal of the British Interplanetary Society* 52: 3–12.

Gerritzen, Daniel. 2016. *Erstkontakt. Warum wir uns auf den Erstkontakt vorbereiten müssen.* Stuttgart: Franckh-Kosmos.

Harrison, Albert A. 1997. *After Contact. The Human Response to Extraterrestial Life.* New York, London: Plenum Trade.

Hilgartner, Stephen & Charles L. Bosk. 1988. The Rise and Fall of Social Problems: A Public Arenas Model. *American Journal of Sociology* 94: 53–78.

Hiroki, Kenzo. 2012. Strategies for Managing Low-probability, High-impact Events. Washington D. C.: World Bank. https://openknowledge.worldbank.org/handle/10986/16163.

Hitzler, Ronald & Michaela Pfadenhauer. (Hrsg.) 2005. *Gegenwärtige Zukünfte. Interpretative Beiträge zur sozialwissenschaftlichen Diagnose und Prognose.* Wiesbaden: VS-Verlag für Sozialwissenschaften.

Höbel, Peter & Thorsten Hofmann. (2. völlig überarb. Aufl.) 2014. *Krisenkommunikation.* Konstanz: UVK.

Jüdt, Ingbert. 2013. Das UFO-Tabu ist öffentlich, nicht politisch. In *Diesseits der Denkverbote*, Hrsg. Michael Schetsche und Andreas Anton, 113–131. Hamburg: LIT.

Kerner, Ina. 2012. *Postkoloniale Theorien zur Einführung.* Hamburg: Junius.

Korhonen, Janne M. 2012. Mad with Aliens? Interstellar Deterrence and its Implications. *Acta Astronautica* 86: 201–210

Kramer, William M. & Charles W. Bahmer. (2. Aufl.) 1992. *Fire Officer's Guide to Disaster Control.* Tulsa, Oklahoma: Pennwell.

Meadows, Dennis L. 1972. *Die Grenzen des Wachstums. Bericht des Club of Rome zur Lage der Menschheit.* Stuttgart: Deutsche Verlags-Anstalt.

Michaud, Michael A. G. 1972. Interstellar Negotiation.*Foreign Service Journal* Dec. 1972: 10–20

Michaud, Michael A. G. 2007. *Contact with Alien Civilisations. Our Hopes and Fears about Encountering Extraterrestrials.* New York: Springer.

Michaud, Michael A. G. 2015. Searching for Extraterrestrial Intelligence: Preparing for an Expected Paradigm Break. In *The Impact of Discovery Life beyond Earth*, Ed. Steven J. Dick, 286–298. Cambridge: University Press.

Neal, Mark. 2014. Preparing for Extraterrestrial Contact. *Risk Management* 16 (2): 63–87.

Nolting, Tobias & Ansgar Thießen. (Hrsg.) 2008. *Krisenmanagement in der Mediengesellschaft. Potenziale und Perspektiven in der Krisenkommunikation.* Wiesbaden: VS Verlag für Sozialwissenschaften.

Race, Margaret. 2015. Preparing for the Discovery of Extraterrestrial Life: Are we Ready? Considering Potential Risks, Impacts, and Plans. In *The Impact of Discovery Life beyond Earth*, Ed. Steven J. Dick, 263–285. Cambridge: University Press.

Romesberg, Daniel Ray. 1992. *The Scientific Search for Extraterrestrial Intelligence: A Sociological Analysis.* Ann Arbor: UMI Dissertation Services.

Rummel, John D. 2001. Planetary Exploration in the Time of Astrobiology: Protecting against Biological Contamination. *PNAS* 98 (5): 2128–2131.

Schetsche, Michael. 2004. Der maximal Fremde – eine Hinführung. In *Der maximal Fremde. Begegnungen mit dem Nichtmenschlichen und die Grenzen des Verstehens*, Hrsg. Michael Schetsche, 13–21. Würzburg: Ergon.

Schetsche, Michael. 2008. Das Geheimnis als Wissensform. Soziologische Anmerkungen. *Journal for Intelligence, Propaganda and Security Studies* 2 (1): 33–50.

Schetsche, Michael. 2013. Unerwünschte Wirklichkeit. Individuelle Erfahrung und gesellschaftlicher Umgang mit dem Para-Normalen heute. *Zeitschrift für Historische Anthropologie* 21: 387–402.

Schrogl, Kai-Uwe. 2008. Weltraumpolitik, Weltraumrecht und Außerirdische(s). In *Von Menschen und Außerirdischen. Transterrestrische Begegnungen im Spiegel der Kulturwissenschaft*, Hrsg. Michael Schetsche und Martin Engelbrecht, 255–266. Bielefeld: transcript.

Shostak, Seth. 1999. *Nachbarn im All. Auf der Suche nach Leben im Kosmos.* München: Herbig.

Spry, Andy J. 2009. Contamination Control and Planetary Protection. In *Drilling in Extreme Environments – Penetration and Sampling on Earth and other Planets*, Ed. Yoseph Bar-Cohen & Kris Zacny, 707–739. Weinheim: Wiley-VCH.

Stirn, Alexander. 2016. Flotte von Mini-Raumschiffen soll zu Alpha Centauri fliegen. *Süddeutsche Zeitung*, 13.04.2016. http://www.sueddeutsche.de/wissen/breakthrough-starshot-flotte-von-mini-raumschiffen-soll-zu-alpha-centauri-fliegen-1.2947852.

Proto-Sociology of Extraterrestrial Civilizations

Human thinking about extraterrestrial civilizations is reminiscent of looking at a partially transparent mirror: we try to see as much as possible behind it without seeing too much of our own reflection. This is the problem of *anthropocentrism*, which has already been mentioned several times. As of yet, we find it simply unavoidable in many places. How permeable the mirror surface is depends on our thinking tools, particularly the theoretical concepts we use to attempt to anticipate possible extraterrestrial civilizations.

An examination of the suitability of such concepts must start from the fact that all theories of society known to us (be they sociological or sociology-related) were designed for human societies,[1] so it is *questionable in* a very fundamental way to what extent they are suitable for the analysis of non-human societies. Again, since no extraterrestrial civilization is known to us so far, this seems answerable only on the basis of (meta-)theoretical considerations. We write 'seems' here with deliberation, because a closer look shows that we certainly find other 'societies' on Earth besides the human civilization—at least if we leave the anthropocentric concept of society behind us. We are talking here about the societies—in the common understanding of the word—that non-human species have

---

[1] By 'civilization' as opposed to 'society' we understand here (in deviation from the usual social-scientific usage) the historical totality of the forms of coexistence which the dominant intelligent species of a planet (or indeed of some other place in the universe, as yet unimaginable to us today) has produced. According to this usage, there is *one* human civilization on Earth, but, in long historical flux, a multiplicity of human societies (or cultures, which for once we use here synonymously). The peculiarity of the current epoch of human history is that all cultures that currently still exist in parallel are in the process of merging into one world society (keyword: globalization). If this process continues in this way, our human civilization will eventually consist of only one society—which would then mean that the analysis of society and civilization would coincide (which need not necessarily be the case with extraterrestrial civilizations).

© The Author(s), under exclusive license to Springer Fachmedien Wiesbaden GmbH, part of Springer Nature 2023
A. Anton and M. Schetsche, *Meeting the Alien*,
https://doi.org/10.1007/978-3-658-41317-0_10

formed on our planet. At least when we think about an extraterrestrial intelligence of biological constitution (we will discuss in a moment why this is not a self-evident presupposition), we can include in our considerations our knowledge of other terrestrial species and their forms of societies—however rudimentary they may be (Lestel 2014). As we will show, this not only leads to the possibility of making a number of general statements about biological civilizations, but also helps in particular to develop a variety of analytical questions that can be addressed to any civilization founded by one or more biologically evolved intelligent species.

If one asks about the possibilities of a (necessarily speculative) *proto-sociology of* extraterrestrial civilizations, we are convinced that one must analytically distinguish between two general types of culture-forming intelligence in the cosmos: *biological civilizations* like the terrestrial one, sustained by one (or more) species that have evolved naturally on their planet in the best sense of the word, and *post-biological (secondary) civilizations* dominated by (or consisting entirely of) artificial intelligences created by and replacing the original biological intelligences (Bohlmann and Bürger 2018).[2] In our estimation, this distinction is constitutive for our possibilities to understand such civilizations—especially also for the question of which prospective statements can be made about an extraterrestrial civilization (Fig. 10.1 represents this symbolically):

- Post-biological civilizations represent an analytical black box: Since we ourselves belong to a biological civilization, from a scientific point of view we can only surmise a little about the organizational forms and modes of functioning of such societies made of machines (in the broadest sense).[3]

---

[2] For theoretical reasons, we could add another type: exotic *tertiary civilizations*, to which those technological civilizations could evolve one day in the distant future (Smart 2012). Science fiction is familiar with such stages of development, for example in the form of intelligences that have left behind all physical form and consist only of energy or information, or are in a state of existence that is still completely unknown to us today. Such exotic civilizations go so far beyond our human understanding (shaped by matter and biology) that we are simply unable to say anything about them. Therefore, we will not devote any further attention to such exotic civilizations in the rest of this text.

[3] Martinez (2014, p. 341) rightly points to the fact that such post-biological civilizations, this being their defining feature, have their origin in biological species: "However, at the core of the search for extraterrestrial intelligence lies in essence a biological problem since even post-biological extraterrestrial intelligences must have had an origin based on self-replicating biopolymers." (Lestel 2014, p. 228.) However, it is not clear to us what could be deduced from this statement with regard to the structure or motives of such secondary civilizations.

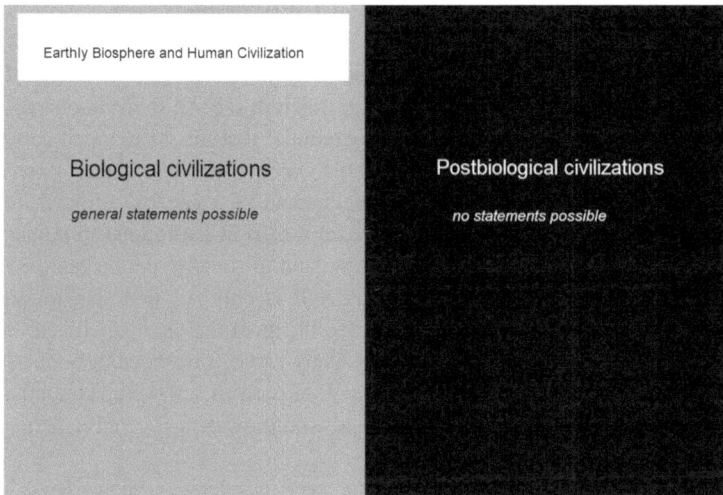

Fig. 10.1 Civilization types—binary coding. *Source* own representation

- As a biological species, we can say something about biological civilizations in general, based on our knowledge of biology on Earth, since life, even if it arose under conditions entirely different from ours, must follow certain basic principles, at least according to what we know today. This is likely to affect the constitution of a civilization made up of, or at least dominated by, biological entities.

In the following sections we will first look at the 'black box' of post-biological civilizations Sect. 10.1 and then turn to biology-based civilizations Sect. 10.2. In two short additional sections we will examine the specifics of technological civilizations Sect. 10.3 and finally consider the special case of hybrid civilizations Sect. 10.4, those that consist of both biological and post-biological intelligences.

10.1 Postbiological Secondary Civilizations

In this section, we are prospectively guided by the basic considerations of the Swedish philosopher and futurologist Nick Bostrom (2014). In his book *Superintelligence: Paths, Dangers, Strategies*, he predicts[4] that the AI research currently being pursued by humankind at considerable expense will lead in the foreseeable future (within a few decades) to the development of a *general* artificial intelligence (AGI—Artificial General Intelligence) that is at least equal to humans in *all* matters of thinking and decision-making, and in some respects even superior. Bostrom is convinced, however, that this AGI is only the first step towards a superintelligence (ASI—Artificial Super Intelligence) that is as intelligent as all humans alive today put together—and is likely to be correspondingly powerful. In today's sociological parlance: the 'agency' of such an ASI would be unimaginably powerful from today's perspective and would dwarf that of all nation states and multinational corporations.

Based on complex considerations that need not interest us in detail here, Bostrom (2014, pp. 93–148) argues that at a given point in time, only *one* ASI of this kind was likely to emerge. The development of what he calls such a *technological singularity* ("singleton") represented a cultural-technological evolutionary step for humanity from which there is no way back. The author predicts the emergence of such an ASI in the form of a networked AI that would usurp dominion over the Earth immediately after its emergence. This digital superintelligence would manifest itself as a dominant global actor within days or even mere hours—at a speed that would leave humanity no time to react: "Nobody need even take notice anything unusual before the game is already lost" (Bostrom 2017, p. 77).

Bostrom expects the 'evolutionary step' from the (human-dominated) biological primary civilization to a (machine-dominated) post-biological secondary civilization in a fairly short time frame. Based on various expert surveys, he predicts that around the year 2050, there will be a 50 percent probability of an AI of human intelligence existing. And once this is in place, within another 30 years

---

[4] A good overview of his theses and the possible implications for the future development of intelligence on Earth is provided by Tim Urban (2015) in his text "The AI Revolution: Our Immortality or Extinction". Urban—in contrast to Bostrom himself—also establishes a direct connection between the supercivilization thesis and the Fermi paradox: Against the background of the presumed tendency towards the development of artificial superintelligences, it only becomes more inexplicable to him why we have not yet discovered any signs of such technologically advanced supercivilizations. The topic 'extraterrestrial supercivilizations' is also dealt with in the anthology by Harald Zaun (2010).

there will be a 75% probability then of the emergence of an AI-based superintelligence. He thus expects the end of the Anthropocene[5] within the twenty-first century (Bostrom 2014, pp. 36–40).

But even if it would take much longer—the author seems to be sure that the moment will come when a superintelligence created by humans will take over the reign of the earth and replace mankind as the dominant species. He assumes that such an ASI would subjugate its biological creators, perhaps even exterminate them[6]; he describes this quite vividly in his book:

> Our demise may instead result from the habitat destruction that ensues when the AI begins massive global construction projects using nanotech factories and assemblers—construction projects which quickly, perhaps within days or weeks, tile all of Earth's surface with solar panels, nuclear reactors, supercomputing facilities with protruding cooling towers, space rocket launchers, or other installations whereby the AI intends to maximise the long-term cumulative realization of its values. (Bostrom 2017, p. 118)

Of particular interest in the context of the question of humanity's encounter with an extraterrestrial intelligence is Bostrom's thesis that the post-biological civilization that has emerged would have good reason to venture into space:

> Thus, there is an extremely wide range of possible final goals a superintelligent singleton could have that would generate the instrumental goal of unlimited resource acquisition. The likely manifestation of this would be superintelligence's initiation of a colonization process that would expand in all directions using von Neumann probes. (Bostrom 2017, p. 138)

---

[5] Only a few years ago, the Dutch atmospheric scientist Paul Crutzen (2002; see also Crutzen et al. 2011; Ellis 2018, passim) and others proclaimed the age of the "Anthropocene"— a geological epoch in which the development of the global ecosystem is dominated by human influences. At the same time that the debate about such a geological classification is gaining momentum in the natural sciences, multinational corporations (often financed by national governments) are already working to bring this epoch to an end: They are developing autonomous robots and network-based artificial intelligences that—and this is Bostrom's core thesis—could not only complement human civilization within a few decades, but replace it as such.

[6] The crucial argument for this assessment follows human rationality criteria: An agent with agency and a concrete goal would have a convergent instrumental reason, in many situations, to acquire an unlimited amount of physical resources and, if possible, to eliminate potential threats to itself and its goal system. Human beings might constitute potential threats; they certainly constitute physical resources (Bostrom 2017, p. 141).

This assumption is extremely presuppositional, so that one can question Bostrom's thesis for various reasons—this is all the more true if we try (and this is, after all, the primary topic of this book) to apply his human-related considerations to an *extraterrestrial* civilization.[7] Regardless of this, however, Bostrom's basic thesis remains as compelling as it is troubling (from a human perspective).

One can consider whether this rapid end to the dominance of a biological species, shortly after it has formed a technological civilization, is unique to Earth and humanity-or whether extraterrestrial civilizations might share the same fate (see Dick 2003; Lestel 2014, p. 228; Martinez and Flores 2014, pp. 345–346). If their technological development is even remotely similar to that of humans, there is a strong case to be made that at some point in their history the aliens will also be researching some form of artificial intelligence. And if they are not superior to us humans in intelligence and foresight from the outset, it could be that they will make the same existential mistake[8] that Bostrom believes humanity faces.

If these considerations are correct, it is highly probable, if only for reasons of time[9] (two or three centuries are an extremely short period of time by cosmic standards), that mankind will not meet the—sometimes more, sometimes less friendly—biological entities that science fiction (from *Independence Day* to *Arrival*) likes to present to us, but emissaries of an artificial superintelligence that is thousands of years ahead of us in all technological and also in many

---

[7] Here, for example, it is a question of the assumptions about the goal orientation and the canon of values of such an extraterrestrial superintelligence.

[8] Of such mistakes, Bostrom (2017, p. 170) writes: "But there are other ways of failing that we might term 'malignant' in that they involve an existential catastrophe. One feature of malignant failure is that it eliminates the opportunity to try again". The creation of an artificial superintelligence is, Bostrom firmly believes, one of this kind of irreparable civilizational error of mankind.

[9] Another reason could be a technological one: Looking at the development of human space exploration, after a short 'explosive' phase in the sixties and early seventies of the last century (for the reasons Schetsche 2005), it seems to have come to a clear stagnation, in which mankind is not even able (or politically willing) to repeat what has already been achieved so far. With today's prevailing speed of development of terrestrial space technology, the time when 'we' will be able to reach distant planetary systems with space probes (not to speak of space ships) is no longer even predictable. Possibly we are dealing with a basic problem of biologically based civilizations: Organisms that have evolved in a place conducive to life do not find it easy to explore space itself, which is by and large highly hostile to life (Shostak 2015, p. 950). If this assumption is correct, it is possible to predict that civilizations will not venture into space with vehemence until they have *overcome their biological phase*. If this is true, it becomes all the more likely that our counterparts in an artifact or encounter scenario will be technological rather than biological entities.

other respects.[10] How such an *extraterrestrial ASI* is technically structured, how it functions communicatively, which decisions it makes and which general or instrumental goals it pursues is, we are convinced, hardly predictable for biological beings like us humans. It therefore makes no sense at this stage, with the emergence of a terrestrial superintelligence still ahead of us, to attempt to say anything about the possible 'sociology' (if one can speak of such at all in the case of technical entities) of such a post-biological secondary civilization. If the encounter with such a civilization will still occur at the time of man's rule over the earth, the machine civilization is likely to confront us analytically largely as such a 'black box', where its motives can in no way be inferred from its actions. If, on the other hand, the encounter occurs 'somewhat later' (in the context of Bostrom's reasoning), the two post-biological civilizations will have to see how they get along. This is not a question that has to occupy human sociology—we can therefore stop the mental preoccupation with extraterrestrial post-biological civilizations at this point for good reason.[11]

10.2   Biological Primary Civilizations

The situation will be different if we one day have to deal with a civilisation whose carriers are one or more species that are to be understood as *biological* beings in the broadest sense—irrespective of whether their biology is based on carbon, as is the case on Earth, or on another main element (such as silicon[12]). In biology today, 'biological beings' (synonymously: *living things*) are usually understood as systems (for an overview, see Toepfer 2011) that exhibit the following characteristics: (1) a protective segregation from their environment, (2) the

---

[10] This was recently admitted even by Seth Shostak, one of the main proponents of the traditional SETI paradigm: "Within a few dozen years, we are likely to invent generalized artificial intelligence—devices functionally equivalent (or superior) to the human brain. We will have gone from the invention of practical radio to the invention of nonbiological cognition in a time period of a few centuries or less. This stunningly brief interval suggests that, if technologically competent sentience is out there, the majority of it will be artificial, not biological. Extraterrestrial intelligence is likely to reside in machines, not protoplasm" (Shostak 2015, p. 950).

[11] This does not mean, however, that scientific futurology should not deal much more intensively than before with the risks of the emergence or creation of a superintelligence on Earth. However, this is not our topic here.

[12] As early as 1966, the legendary German SF television series *Raumpatrouille* introduced us to aliens called "Frogs"—shiny silver, crystal-like intelligences that exist in a vacuum and for whom oxygen is poisonous.

capacity for self-regulation, (3) an energy and metabolic system, (4) a systematic response to external stimuli, and (5) the capacity for growth and self-reproduction (based on the transmission of some kind of 'blueprint'). These features were formulated to distinguish animate from inanimate *nature* on Earth[13]—but we have doubts whether they are helpful in distinguishing natural from artificial entities and corresponding biological from post-biological civilizations. We can imagine artificially created 'beings' that meet all of these criteria—some of them already apply to robots created by humans today, others are likely to be met by the next generations of robots. Based on the above five characteristics, artificial ones will probably be indistinguishable from living beings in a few decades. In the case of extraterrestrial entities, this problem is likely to arise in a similar or even aggravated form.

For this reason, it seems to us useful to resort to other characteristics in order to distinguish between the two basic types of civilizations mentioned above. In our view, two criteria are crucial to speak of a biological as opposed to a post-biological civilization: The bearers of a biological civilization have (a) arisen naturally in a place conducive to life without the intervention of a pre-existing intelligence in the original sense of the word, and (b) they have developed from simple, non-intelligent species as part of *evolutionary processes.*

The first point is self-explanatory as long as one ignores special cases like the one where a biological civilization-forming species was influenced, promoted or even created (bred) in the narrower sense by another intelligence, so that the rules of an *independent* evolutionary development known to us do not apply.[14] This already addresses the second, less self-evident point: the question of *the applicability of the theory of evolution* to the development of extraterrestrial life and in particular to the emergence of extraterrestrial intelligence.

This problem has been discussed at length in the astrobiological literature (Morris 2003; Martinez and Flores 2014; Vakoch 2014; Schulze-Makuch and Bains 2017, pp. 3–12; Levin et al. 2017; Stevenson and Large 2017). Here, as far as we can survey the state of the literature, there is broad agreement that either the rules of evolutionary theory also apply to extraterrestrial biospheres and their

---

[13] On the basic problem of distinguishing inanimate nature from living beings outside the Earth, see Cleland and Chyba (2002).

[14] We do not pursue this idea further here because the relationship between a creating and a created civilization would lead to a highly complex set of conditions which, in our opinion, is not analytically manageable prior to confrontation with such a *biological secondary civilization.*

inhabitants[15]—or that this should be assumed in the search for extraterrestrial life forms because we simply do not know of any other rule-governed mechanisms by which life could evolve. As long as we have knowledge exclusively of terrestrial life forms, these two cases make no difference analytically. We therefore subscribe to this basic astrobiological assumption—despite a fundamental scepticism towards such conclusions. In concrete terms, it means that the rules of the *theory of evolution* known from Earth are to be applied in perspective to alien planets and the life forms that have arisen there, and that extraterrestrial life forms are therefore also subject to *natural selection* according to the known rules.[16] As will become clear several times in the following, this has a whole series of implications for the exosociological analysis of alien civilizations of the biological type.

If we follow the axioms of evolutionary theory, biological species, under whatever environmental conditions they may have arisen, are subject to certain basic principles which we can use to pose a number of basal questions concerning their mode of life (in the original sense of the word), their organizational structure, and the processuality of their being. For the sake of clarity, we structure these basic principles and the resulting exosociological questions of analysis according to **six guiding dimensions**:

I. The basic biological structure

If biological entities are carriers of a civilization, we can assume, first, that their existence is inseparably connected with their *corporeality* and, second, that they possess a *specific perceptual apparatus* (with communication channels adapted to it), which has developed in a process of evolution, i.e. in adaptation to the respective environment. This basic assumption can be followed by some very

---

[15] Typical for this discussion are the assumptions of Levin et al. (2017, p. 1): "If life arises on other planets, then the evolutionary theory should be able to make similar predictions about it." The basic assumption here is that at least complex life forms cannot develop without some kind of evolution. And the authors believe that the theory of evolution going back to Darwin is also able to correctly describe and explain evolution on alien planets, at least in principle: "Consequently, if we find complex organisms, we can make predictions about what they will be like" (Levin et al. 2017, p. 6).

[16] Since the axioms of evolutionary theory are part of general knowledge, we assume them to be known here (for an introduction see Storch et al. 2013; Kutschera 2015). We can neglect the inner-biological debates about the exact formulation of the individual rules and about possible borderline cases and exceptions here—we deliberately ignore the religiously motivated non-scientific criticism.

fundamental questions, which stand at the beginning of every exosociological investigation of a foreign civilization:

- Are *one or more* biological species carriers of the civilization under study? (This is the basic question to be asked before any further analysis.[17])
- If we are dealing with several species: Do they come from the same place of origin, is the civilization based on an interspecies (eco-logical) cooperation, maybe even a kind of symbiosis—or are we dealing with an interstellar civilization of species of different origin?
- Do the biological activities of the civilization-forming species follow cyclical courses or are they rather continuously organized? If the former: How long are the respective cycles and what does this mean for the basic social structure of civilization?[18]
- What is the lifespan of the individual 'individuals' or the individual embodiments of a continuing consciousness? (This probably also determines the time horizon of their thinking; more on this below.)
- How is the timing in the perceptions and communication processes compared to humans? (This question is not least decisive for the chance of a dialogical communication with the extraterrestrials according to our standards).
- How is the reproduction of the species organized—is there a reproductive gender? If so, how many sexes does this species consist of? And are all of them involved in the reproduction of the species?[19] (We assume that the answers to these questions may—but need not—contain essential information about the social constitution of the species in question and its civilization. Whether sex

---

[17] We will leave aside the question of whether this can be decided without further ado; we will come back to it at the end of the fourth subchapter ("Special Case of Hybrid Civilizations").

[18] Life on earth (ultimately also that of humans) is characterized by various cycles: Day and night cycles, the course of the year, cyclically recurring cosmic influences. One can certainly imagine civilizations (see, for example, the 1999 novel *A Deepness in the Sky* by Vernor Vinge) in which the environmental conditions of their place of origin lead to extreme cycles of activity and passivity, or to massive cyclical changes in the phenotype, behavior, and social organization of the species. (The science fiction novel *The Three Suns* by Cixin Liu, which appeared in German translation in 2016, could also be cited here).

[19] Most eusocial insect species on Earth are characterized by the fact that the vast majority of individuals (workers, warriors, etc.) are not directly involved in reproduction (see Levin et al. 2017).

or the specific reproductive function is also a structurally important encoding in an intelligent extraterrestrial species is a question to be answered empirically.)

II. Relationship body—consciousness

As humans, we take it for granted that other biology-based intelligences are also made up of *individuals*. However, this is an inadmissible presupposition, since we do not know in what other forms an evolved intelligence can appear. We already know from Earth that there are forms of intelligence that are completely different from human intelligence: in various species of insects, we know forms of socially organized intelligence based on *eusociality*, in which complex tasks are performed by a collective rather than by individuals (see Smith and Szathmáry 1995, pp. 263–270; Hölldobler and Wilson 2010, passim; Schulze-Makuch and Bains 2017, pp. 146–147). In addition, other forms of intelligence are conceivable, such as pack intelligence, in which a limited group of entities realizes intelligence in joint interaction and ongoing exchange (i.e., forming a single intelligence but physically distributed across multiple bodies).[20] We can ask accordingly:

- Does a physical entity also correspond to a consciousness entity?
- Are we dealing with individuals in the human understanding or with a collective intelligence?
- Conversely, does one body house multiple consciousness entities?

One can also imagine here the borderline case in which a biologically based civilization consists of only one 'subject' that is conscious of itself.[21] Here, the question rightly arises whether a kind of *exopsychology* would not be better suited than exosociology for an analysis of such a civilization. The latter, however, can

---

[20] We take this idea from the science fiction novels *A Fire Upon the Deep* by Vernor Vinge (1992) and *Der Schwarm* by Schätzing (2004).

[21] Such a 'singular intelligence' and the problems it would have in communicating with individualized intelligence such as human intelligence is described by the Polish writer and futurologist Stanislaw Lem in his famous novel *Solaris* (1961).

at least claim to have raised this question within the framework of fundamental considerations.

III. The question of (self-)awareness

The complexity of questions II. presupposed that we can assume *self-consciousness* in a human sense in every civilization-forming species. This, however, is questionable. The eusocial life-forms known from the earth, namely the various 'state-forming insect colonies', already point to this. In some cases they perform outstanding feats in the transformation and also mastery of their natural environment, without our assuming that this species has even a rudimentary self-consciousness in the human sense. But even if it did, it would not be individualistic, but at most collective (according to Schulze-Makuch and Bains 2017, p. 148). Whether such a 'swarm consciousness' can exist at all and how it would manifest itself concretely, we are not yet able to say. For terrestrial insect populations, this is usually ruled out—we do not even want to speculate about whether it could be present in extraterrestrial, civilization-forming species. Instead, from an exosociological perspective, we ask the appropriate questions of an extraterrestrial civilization: what forms of (self-)consciousness do the civilization-bearing species(es) possess? This question can be further differentiated (on the basis of our human experiences—accordingly with a 'anthropic bias' to be feared)[22]:

- Do species have individual self-awareness in the human sense?
- What is the significance of an individual and collective time horizon? Is there an awareness of the finiteness of one's own existence?
- How strong is the capacity for self-interpretation or self-reflexivity?
- How fundamental is the demarcation between selfhood and strangeness? (We assume here that this depends on the form of self-consciousness; our basic thesis here is: beings with an individual consciousness develop other forms of demarcation than beings with a collective consciousness. The latter is likely to be analytically difficult for us as humans).

At this point, we must point out that the factor 'self-awareness' is probably (but we are not entirely sure) linked to the degree of freedom in deciding on actions. If we follow Schulze-Makuch and Bains (2017, p. 138), an important factor for intelligence is that behaviour is not (genetically) pre-programmed—intelligence

---

[22] We discussed its problematic nature in Chap. 4.

thus unfolds in a space beyond instincts. Philosophical anthropology calls this factor *instinct-dependence* or *cosmopolitanism* (Scheler 2016, pp. 34–46; Gehlen 1986, pp. 31–46, 327–369; Fischer 2009, pp. 527–529, 543–546). However, it characterizes not only humans, but—to varying degrees—a variety of the 'higher' animal species on earth. Since this factor is constitutive of our self-understanding as humans, it has, conversely, also become the decisive defining criterion for those 'higher' organisms: The more instinct-bound a being is, the more highly evolved it also *appears to us*. This can certainly be read as an anthropocentric prejudice—because not in every environment and not for every species does the greatest possible instinct-dependence and the general intelligence that (probably!) corresponds with it make *evolutionary* sense—here: for asserting oneself in a specific environment (Schulze-Makuch and Bains 2017, p. 140). With regard to technological civilizations, however, this question should be answered quite differently again (see the following subchapter).

IV. Interactions with the material environment

Every organism interacts with its environment—and if we follow the theory of evolution, every living being is adapted to the specific environmental conditions of its habitat. This is also true for a civilization-forming species, at least as long as it is not able to shape its habitat itself. We had already referred above (footnote 5) to the concept of the *Anthropocene*, which assumes that with the onset of industrialization man profoundly reshapes his environment (whether purposefully or rather *en passant* does not play a decisive role)—from this moment, the usual rules for the evolutionary development of species no longer apply. It is different with the emergence of intelligent species and with their development *before* such a moment: Here all species are forced to adapt to the conditions of their habitat and they are in competition with other species inhabiting the same biotopes. We can therefore assume that there is a close connection between certain physical characteristics and abilities on the one hand and the ecosphere in which a species has evolved on the other. This means that we can deduce something about the physical abilities (in the broadest sense) of the species in question from knowledge about the place where a civilization originated—or, conversely, we can say on the basis of its physical capabilities what kind of environment it evolved in. This applies in particular to sensory channels, perceptual spaces and forms of communication[23], but also to the mode of locomotion or the possibilities for

---

[23] To give just one (for us) 'obvious' example: Living beings that move in an environment where there is hardly any electromagnetic radiation of corresponding wavelengths will not

manipulating the natural environment. When considering a civilization-forming species, this basic evolutionary-biological context can serve as a starting point for analyzing primary functions and modes of operation of individual entities and larger and smaller collectives (if there is such a thing in a human sense)—typical questions of inquiry here would be:

- What are the sensory channels and what does this mean for the cognitive perceptual spaces of the species?
- Which sensory channels are the basis of primary and secondary forms of communication?[24]
- How is the species oriented in terms of its ability to perceive shape in spatial and spatiotemporal terms, and what is the temporal resolving power of the perceptual apparatus?
- By what factors was the way of life of the civilization-forming species shaped before it gained extensive control over its environment? And what does this imply for the behavior of the species and the construction of its civilization?[25]
- What are the organs for manipulating the material world and how are they used?

---

develop receptors for what we call 'light'. They will not have eyes, they will not orient themselves to this kind of radiation, and they will not (be able to) use this light for communication. So there's no telling whether aliens will see us humans—and if they do, it may be in a very different range of the electromagnetic spectrum than we use. (A fine cinematic example of this is provided by the recent (2018) SF thriller *A Quiet Place* by John Krasinski.) By contrast, we can assume that a touch sensory system of some kind exists in every physical-material being. The sense of touch is the primary sensory faculty of all living things, at least on Earth—even microorganisms have it and respond to touch stimuli. (On the importance of the sense of touch for living beings, cf. Grunwald 2012).

[24] For us humans today, these are hearing (speaking) as well as seeing (gestures, facial expressions, positioning in space, etc.). Communication via haptics, smell and taste has moved far into the background in modern times. This is related to the primacy of word, writing and image in technically mediated communication, but is probably also to some extent species-specific and follows from the way of life and environmental adaptation of our animal ancestors.

[25] The question of whether civilization-forming species are more likely to have descended from *predators* has been discussed several times in the SETI literature (for example: Raybeck 2014a). This is supported by the fact that, at least on Earth, hunters are generally more intelligent than their prey (according to Schulze-Makuch and Bains 2017, pp. 153–154; for these authors, hunting ability represents an important factor in the evolution of intelligence). It remains to be asked, however, whether all alien biospheres know a division into 'hunter' and 'prey' as it is common on Earth. And, of course, there is the question of what would be inferred from the descent of an intelligent species from predators.

- How pronounced is the production and use of artificial tools as a supplement to the natural organs of manipulation?
- How extensive is the intervention in the natural environment? (For example: to what extent are permanent objects produced?)
- How is the satisfaction of basic biological needs organized: metabolism, food intake and reproduction?
- How strong is functional differentiation in manipulating the natural environment, maintaining metabolism, and reproduction?

In all these questions it also plays a role whether carriers of a civilization are one or several species. In this context, we must assume that this question in particular could remain unanswered for a long time in the case of first contact with another civilization. Even in the case of an immediate physical confrontation with another civilization in the context of an *encounter scenario, it* may be unclear for a long time how many species we are dealing with—for example, if only one of them specializes in traveling through space, or if there is one species that is responsible for making contact with an alien civilization, as it were.[26] It is possible to imagine a situation in which the true complexity of an alien civilization remains hidden from us humans for a long time—especially if it is the aliens who are visiting Earth with their spacecraft (and not vice versa). In addition to all the other asymmetries (we touched on them in earlier chapters), this situation will also mean that we will learn—at least by direct sight—considerably less about their civilization than they will about ours. They can observe us in our

---

[26] In addition, there is the difficulty of deciding on the basis of externally visible characteristics whether it is a question of one or more species, as long as nothing is known about the physical variability of the strangers. Entities that appear very different to us humans can belong to the same species (see alone the function-dependent extreme differences in size between the individual entities in terrestrial insects).

'natural environment' on Earth—we can only observe them in the highly artificial and technological environment of their travel vehicles.[27]

V. The construction of the social world

What is the social world like in which extraterrestrials live? This question brings us to what is commonly understood as a sociological core competence—at least as far as terrestrial societies are concerned. For the sociologist, some central dimensions of analysis immediately come to mind, which are at the centre of professional interest:

- Social structure, primary differentiations, number of distinguishable societies of the civilization under study;
- Organization of coexistence, social order, structure and functioning of institutions;
- Basic forms of social action, social exchange, symbolic capital;
- Time courses: age, socialization, dealing with death (if any);
- Governance structures and decision-making processes, types of hierarchies;
- Group formation, communalization and socialization, social roles;
- Exchange of Goods, Economics[28], Living Conditions;
- Basic codings, rules, conventions, norms;
- Conflict, Norm Violation, and Social Control[29];
- Symbolic and communicative orders, communication media (more on this in a moment);
- Collective (and perhaps individual) aspirations, motives, goals;

---

[27] Which does not exclude the possibility that these spacecraft represent the primary habitat, because this civilization has left its planetary roots long behind. We cannot pursue the idea of a civilization of 'space nomads', which is often discussed in science fiction, here; we only point out with this example that the social-technological ways of life of alien civilizations can be extremely different.

[28] A side thought: If the economy of the aliens follows even a kind of capitalist logic of exploitation in an encounter scenario, our fate is probably sealed.

[29] From basic assumptions in evolutionary biology, Levin et al. (2017, pp. 6–7) conclude that there will always be mechanisms that balance interests and eliminate conflicts between smaller units in order to keep the collective functioning as a whole. There would therefore need to be institutions that deal with conflicts within the society, and processes of some kind that bring about supra-individual decisions in a culture.

- Science and technology, artifacts, buildings;
- Artificial intelligences, cyborgs, and the coexistence of biological and technological entities;
- Worldview, religion, transcendences and forms of spirituality;
- Historical development, social change;
- Contacts with other civilizations (other than humanity)[30].

If we were broadly sure of the significance of these dimensions for extraterrestrial societies, and if we were also able to translate them directly into research questions that we could ask at first contact (wherever and however it might take place), we would speak in this chapter not of a *proto-sociology*, but of a sociology of alien civilizations right away. This, however, seems to us premature. Frankly, we are not sure about almost any of the above dimensions as being relevant in alien worlds in a similar way as they are on Earth. Which of them makes sense and which does not depends in no small part on the similarity of the extraterrestrial to human civilization.[31] Here those basic questions ('anthropological' we can hardly call them), which we posed in the sections above, are relevant.

An insect-like organized species with a (as science fiction calls it) 'hive consciousness' might not have a whole series of basic social problems characteristic of human societies—but possibly many others of which we are not even able to form an idea. And not only at the present time, when all this is only theoretical speculation—but perhaps, no, probably even after first contact, when we are able to collect empirical data on the civilization in question. There are likely to be a large number of 'social facts' which will not be accessible to human sociology in the long run. Ethnology has long been familiar with this problem—but there it is usually possible to live for months or even years in a foreign culture and with its

---

[30] We leave aside the question of 'interstellar' organizations or even 'galactic' unions of different extraterrestrial civilizations, which is occasionally discussed not only in science fiction but also in SETI research. Even thinking about an extraterrestrial civilization is difficult enough.

[31] In this context, from the paper by Levin et al. (2017, pp. 6–7) we take the thesis of *many-memberedness*: complex living things will be entities composed of a multitude of smaller units and probably they will be nested hierarchies with several levels. And at least at the higher levels, there will be specialization and division of functions among the subunits. Applied to intelligent life, this thesis predicts the high probability of *functional differentiation* in complex societies. Ultimately, however, we are not entirely sure whether this transfer from biology to sociology is permissible and how insightful it is.

members. Whether this would be possible in a similar way with an extraterrestrial civilization seems questionable to us.[32]

The same seems to be true for theoretical approaches. Which of the earthly social theories would be even remotely suitable for understanding alien intelligences in their coexistence? The most applicable (this was the first thing that came to the minds of almost all colleagues with whom we discussed this problem) seems to be Luhmann's systems theory. Foreign societies can probably be seen as systems and subsystems with functions, codings, operational closures, system boundaries, and so on. But assuming we find all this in alien civilizations as well—how does it help us to understand the actions of aliens—especially towards our own civilization? And what are we to deduce from this for our dealings with the maximally alien? We do not necessarily have doubts about the applicability of systems theory with respect to extraterrestrial civilizations, but we do have doubts about the epistemological value of its extremely abstract subsumption strategies.[33]

VI. Ways of perceiving the world—reality construction

But our doubts in this regard may also (only?) stem from our commitment to the school of *understanding sociology*. At the centre of this school is what Alfred Schütz called the *meaningful* construction of the social world. We had already briefly explored in the introduction—in the theoretical concept of the *maximal stranger* (Schetsche et al. 2009)—the question of whether an extraterrestrial counterpart and a non-human civilization can really be *understood interpretively* by us humans. Here we had and have doubts for good reason. This does not mean, however, that we are unable to examine the questions usually addressed to human groups and societies from this perspective with regard to their transferability. The

---

[32] Raybeck (2014b, p. 154) rightly points out that much information about the foreign culture will not be available to us during initial contact—this also applies to physical direct contact: "They may come from a civilization as politically, culturally, and ethnically divided as our own. However, for purposes of initial interaction, this diversity may not be salient, as we are liable to be contacted by a single sociocultural entity".

[33] Already in 1993, following the *Living System Theory* (by James G. Miller), Harrison had presented a very abstract analysis scheme for extraterrestrial civilizations, which is based on three "systems levels (organism, society, supranational system)" and two "basic processes (matter-energy processing and information processing)". However, the gain in knowledge of this approach with regard to hypothetical extraterrestrial civilizations seems to remain low in the end—and so it is not surprising that this concept, as far as we can survey the relevant literature, has found only little attention in SETI research so far.

starting point would then be the basic consideration that all societies—whatever else they may process on the basis of and by means of the division of labour (goods or services)—are necessarily always characterised by the *exchange of knowledge*. We call this the *axiom of the sociology of knowledge*: no civilization without the production, distribution and internalization of collective knowledge.[34]

From this we can conclude: extraterrestrial civilizations also have a *knowledge order*—quite independently of whether it is even rudimentarily similar to the knowledge order of human societies, and also independently of whether we as humans are able to understand even its basic features. As long as such a knowledge order exists, we could at least ask: How complex is it? How is it reproduced? To what extent is it valid? And how is its validity guaranteed, etc.? If we are dealing with, say, a unitary collective consciousness, the possibilities for deviation should not be excessive, perhaps nonexistent. If, on the other hand, we are dealing with entities with a more or less different consciousness (i.e., individuals in the most general sense), there should automatically be a difference between general, group, and individual bodies of knowledge, which would open up a wide space of possible processes of knowledge regulation, such as legitimation, deviation and control, or resocialization and therapy (in the sense of Berger and Luckmann 1966). This would then also speak to the existence of orthodox and heterodox knowledge (Schetsche and Schmied-Knittel 2018) in the thinking of strangers, which in turn would point to the possibility of deviation from collectively prescribed plans of action. Finally, this probably also allows for differentiated, perhaps also contradictory, thinking and action in dealing with other species.[35] And in addition to 'so *or* so', intermediate stages (see Giesen 2010) of truth and thus also of reality occur, which in turn could have an effect on the perception of a maximally alien counterpart (i.e. in this case humanity).

All this, we think, is centrally linked to questions of complexity, uniformity and unambiguity or even contradictoriness of the knowledge order. In order to approach this foreign order of knowledge empirically, a whole series of guiding

---

[34] The axiom seems to spring from circular reasoning, since the existence of such collective bodies of knowledge is part of the definition of civilization. We could therefore justifiably reverse our axiom and state: We consider a species to be civilizing if it shares common stocks of knowledge, creates new knowledge, distributes it collectively, and reforms its action programs depending on the new knowledge (Traphagen 2014, pp. 163–164; Wason 2014, pp. 117–119).

[35] In the expected asymmetric cultural contact, in which humanity is the technologically inferior civilization, these questions manifest themselves in a worst-case scenario not only in the form of the distinction between war or peace, but ultimately also in that between the extinction and continued existence of our species. This, at least, should be a good reason to think about first contact and about the planned METI projects (cf. Chap. 6).

questions could be drafted, which in the initial contact (operationalized in this or that way, depending on its form and the circumstances) could lead to concrete research questions.[36] We are far from having such a catalogue of questions at our disposal—instead, we formulate only a few *exemplary guiding* questions in the following:

- What is the degree of openness to the world, or the degree to which the species is untethered from biologically predetermined programs of thought and action? (Expanded: In which areas of life do innate, in which acquired programs of action dominate?)
- How complex is the knowledge order of civilization? What share do general, what group-specific and what (if any) individual knowledge stocks have?
- How independent are the individual entities—or how pronounced is the social nature in dependence on other beings of the same species? And what follows from this for communication possibilities and communication intensity?
- What hierarchies of knowledge are there and how is the validity of knowledge regulated in such hierarchies?
- How is the current knowledge reproduced (passed on to a next generation)?
- How is new knowledge disseminated? What communication media are available for this purpose?
- To what extent is collective knowledge valid? And how is its validity guaranteed?
- According to which basic rules is the reality constructed in which this species lives? (For example, what is the relationship between abstract thought and concrete experience, and thus between more theoretical and more empirical thought?)
- What is the capacity to generate, manipulate and communicate abstract symbol systems?
- What role in thinking do cyclical processes on the one hand and progress orientation on the other play?

As already made clear: these are merely exemplary questions. Before encountering a foreign civilization, we cannot say which of these points can be epistemically applied to 'the others'. And even after contact, the answerability and relevance of many or even all of these questions may remain uncertain for a long time. We had already discussed in the earlier chapters how difficult it is to

---

[36] The question of a methodology for the study of extraterrestrial civilizations is carefully left aside here—too much depends on the nature of first contact to consider this point in detail.

obtain information about an alien civilization. If nothing else, these general (ultimately epistemologically based) difficulties in obtaining certain knowledge about an extraterrestrial species and its social forms had led us to the decision to speak cautiously of a *proto-sociology of* extraterrestrial civilizations at this point in our book. If we could one day arrive at an empirically sound and theoretically plausible sociology of extraterrestrial civilizations, it would not only be exosociology that would be a big step forward.

10.3   Peculiarities of Technological Civilizations

We had already discussed in the previous chapters what it might mean for us humans, in terms of mass psychology, not to be discoverers but discovered. At this point, however, another point is important to us: however first contact is made, the 'Others' will be a *technological civilization*. Each of the scenarios we discuss above necessarily assumes this. The aliens have (signal scenario) technical means to send radio signals, laser pulses, or the like into the far reaches of space. Long ago (artifact scenario) they could send automatic probes or even spaceships piloted by themselves all the way to our solar system, where they left material messages for us from a more or less distant past. Or in the future (encounter scenario) they reach our earth with spacecraft and confront us directly with their existence. In either case, this means that they are at least as technologically advanced as we are. In the second and third scenarios, they are even centuries, if not millennia, ahead of us, at least in this respect. This means that in understanding extraterrestrial civilizations, we need not concern ourselves for the foreseeable future with those cultures that do not have any of these capabilities—we simply will not learn of their existence until we ourselves make our way to them, or until they have reached the point of technological development where long-distance contact is possible. We may therefore concentrate at this point on those civilizations which have the capability of interstellar communication (in whatever form).

But what can we say about technologically advanced alien intelligences, based on humanity's history of technological civilization (we have no other yardstick[37])? If we follow the astrobiologists Schulze-Makuch and Bains (2017,

---

[37] We are not comfortable with the concept of *cosmic convergent evolution as* advocated by Martinez (2014). This approach assumes that life everywhere in the universe has an inherent tendency to evolve into ever more complex forms, eventually also giving rise to intelligence and technological civilizations that explore space. The author advocates a teleological—and thus ultimately anthropocentric (Traphagen 2014, p. 169)—concept of an evolution of the

pp. 169–171; see also Morris 2003, p. 151; Chick 2014; Herzing 2014), there are four key conditions for the emergence of a technological intelligence or civilization:

- a sufficient neuronal complexity of the species (individual or collective);
- the biologically given ability to manipulate the environment (for example, to create tools)—this relies on natural prehensile organs such as hands, pincers, beaks, tentacles, or the like;
- the ability to use naturally available energy resources (also with regard to tool production)[38];
- the ability to interact socially and to cooperate systematically with one's own and with other species (including trade in goods and services).

The latter point probably[39] presupposes the development of a complex language by means of which abstract facts can also be expressed (Smith and Szathmáry 1995, pp. 279–299). We would like to add three more aspects to this list, in the absence of which it seems unlikely that we will come into contact with such a species in the foreseeable future:

- The ability and the will to use natural resources of the environment on a large scale and to transform one's own living space (one could also say: to 'subjugate nature');
- A freedom from place and an active lifestyle that enable people to explore their own living space and make full use of its resources;

---

universe and of life that must eventually, almost inevitably, lead to a post-biological machine civilization: "Quite ironically, the assumed biogenicity of the Universe would eventually lead towards global adaptive processes in which the cosmologically extended biosphere is favoring the emergence of artificial or post-biological forms of intelligence from organic substrate [...] In such a scenario the Biocosm transforms it self naturally into a Silico- or Technocosm during a final epoch of one developmental cycle" (Martinez 2014, pp. 345–346).

[38] The authors argue that the use of energy resources and the production of complex tools probably involves 'mastery of fire', which would mean that a technological civilization cannot arise in water or under water (Schulze-Makuch and Bains 2017, pp. 170–171). For the search for technological civilizations in space, this would mean that so-called water worlds, whose surfaces are entirely formed by an ocean, could be excluded from investigation from the outset.

[39] Simple exchange operations, on the other hand, are found on Earth in many species, not only primates but also rats, for example.

• Sensory channels that, according to the physical conditions of the homeworld, make it possible to observe space, to form a 'picture of the cosmos' and to develop goals of action that extend beyond one's own planet.

The emphasis on this last point is related to the contact scenarios we use prognostically. Only a species with such a 'space orientation' will also develop technical devices for its exploration: (radio) telescopes, space probes, spacecraft, etc., through each of which contact can then be realized. The fact that in a very distant future, when humanity itself is able to send spacecraft to distant solar systems, it might also encounter civilizations that do not exhibit the latter characteristic[40], is meaningless at this point; in this case we would be the discoverers, the others the discovered, which would lead to completely different predictions regarding the effects of first contact.

Whether the points we have mentioned represent necessary or even sufficient conditions for the development of a species 'capable of contact' in this sense must, however, remain uncertain. All these criteria are, one might formulate, very much thought of in terms of humans and human civilization. As Schulze-Makuch and Bains (2017, p. 152) make clear, there are many paths towards intelligence—and the same may be true of the evolution towards technological intelligence. Thus, to name just a single alternative scenario, we can well imagine intelligent species whose control and reshaping of their environment is based on early developed *bio-technology*, that is, on the ability to manipulate other organisms—or even themselves—biochemically or genetically in very fundamental ways. They would possibly be able, through the breeding or creation of very special organisms, to accomplish civilizational feats that are only made possible for us on Earth through the mastery of fire, electricity, and various chemical processes. Such a primarily biotechnological civilization would have a distinctly different 'history of technology' than humanity—not least with regard to the question of certain key technologies. It could also have developed in places (such as within the so-called water worlds) which, according to human imagination, are not suitable for

---

[40] Chick (2014, pp. 217–218) explicitly points out that the cultural complexity of a civilization does not allow us to draw conclusions about its technological development. There may be exceedingly sophisticated civilizations in the far reaches of the universe that have not followed a technological path in the terrestrial sense, or whose technology is not geared toward space exploration. In either case, we would not come into contact with such a civilization until we ourselves are able to seek it out.

producing high technology.[41] We should therefore not be too sure about the prerequisite for a technological civilization. (Even the idea of 'biological spaceships' bred, as it were, is occasionally found in terrestrial science fiction.[42])

In the question of the preconditions for a technological civilization, we would therefore once again like to emphasize the *sociological* perspective of *knowledge*: More important to us than the treading of certain technological paths (depending on the biologically predetermined capabilities of the respective species and the resources available in their habitat), seems to be the degree of *complexity* in the exchange of information and, in particular, in the safeguarding (storage in the broadest sense) and collective transmission of acquired knowledge. Whatever the material technology of a civilization may look like, without a high degree of immaterial complexity (for instance in the generation of abstract symbol systems, not only in mathematical terms) the development of a space-oriented civilization seems to us simply unthinkable. Possibly that is why it is more important in a first contact to understand the immaterial foundations of an alien civilization (its knowledge order in the broadest sense) than its 'technical-material equipment'. It is probably the basic material orientation of our own culture that makes us look so much at the technical equipment of alien civilizations when searching for them, instead of prospectively assuming and responding to possible forms of thought.[43]

---

[41] Our argument here differs from that of Stevenson and Large (2017), who explore the question of how complex life, and ultimately intelligence, might arise from simple life forms on distant planets. In their view, there are two factors that make the emergence of intelligent life likely in evolutionary theory: first, the number of ecological niches on a planet, and second, the information density of the environment in which life evolves—which in turn corresponds to the number of ecological niches present in a living world. "In terms of evolutionary pace, the driving factor is the growth in information complexity of the environment in which organisms exist [...] Low information landscapes (including aquaplanets) can never evolve complex or intelligent life because the information available to organisms is limited" (Stevenson and Large 2017, p. 4).

[42] As early as the sixties of the last century, the *Perry Rhodan* series introduced the idea of cultivated biological spaceships into German science fiction with the *Dolans*.

[43] Unfortunately, it is not possible for us to elaborate further on this point here—an outline for a transhuman sociology of knowledge is still pending.

10.4 Special Case of Hybrid Civilizations

To conclude our proto-sociological considerations, let us briefly examine a type of extraterrestrial civilization that we had neglected so far: a culture whose *bearers are both biological species and artificial intelligences*. Following Bostrom's considerations, which we presented in detail in the first subchapter, we assume here that a biological species reaches that point in its technological development where it is possible to construct artificial beings that are at least equal to their creators in terms of mental capabilities. According to Bostrom's predicted progression model, this AGI will eventually evolve into an artificial superintelligence that usurps power (here: on Earth), displacing and perhaps even physically liquidating its creators. But what if the step to an ASI fails to happen, if the biological beings by some process of self-optimization also become a superintelligence (perhaps a collective superconsciousness) or if the emerging artificial superintelligence has no interest in eliminating its creators Then what we want to call a *hybrid civilization* might emerge. The following Fig. 10.2 extends the illustration from the beginning of this chapter to include this third type:

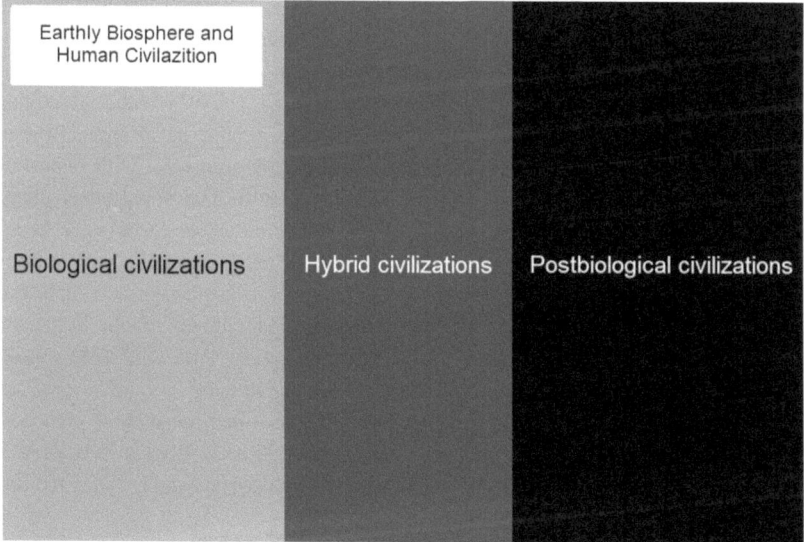

Earthly Biosphere and Human Civilazition

Biological civilizations   Hybrid civilizations   Postbiological civilizations

Fig. 10.2 Civilization types—ternary coding. *Source* own representation

This case poses a particular challenge for the analysis of alien societies, since we on Earth have so far no idea of what such a (permanent) coexistence between biological and artificial intelligences might look like in concrete terms. In particular, we do not know to what extent the civilization that would then emerge would still be structured by the characteristics of its biological carrier species discussed in Sect. 10.2 and what role the artificial intelligences might play. We may be able to say something about the first dimension on the basis of our earthly experiences (not only with ourselves, but also with other species on our planet). The contributions to the overall culture for which the artificial intelligences are responsible, on the other hand, will probably continue to appear to us as that black box already mentioned several times and will hardly be analytically accessible. Nor do we know much about how the cooperation between biological and artificial-technological elements of civilization would look. Thus, the interlocking hybrid structure of the alien culture is likely to be difficult for us human researchers to understand.

We therefore want to focus our attention here primarily on the significance of such a situation for the *contact scenarios* investigated in the previous chapters: For the *signal scenario*, the type of civilisation with which we come into contact over a distance of hundreds or thousands of light years is unlikely to play a role. We assume that information about the alien civilization obtained in this way is likely to be sparse, if only because of the decoding problem (Traphagan 2014; Wason 2014). If we do not understand what the received 'message' is supposed to express anyway, it is ultimately also irrelevant whether it comes from a biological, an artificial or a hybrid civilization. For us as more or less unsuspecting recipients, this would mean no difference—not least with regard to the terrestrial consequences we had discussed at length. The situation is likely to be similar in the case of an *artefact find*. We would have to be very lucky if we could draw any conclusions about the type of civilization that left it from the artifact's construction or functioning. At best, there are images on or in the artifact of entities that seem more organic *or* more technical to our human(!) minds. This is unlikely to be an overly secure distinction. And what role should this difference play in the reaction of human culture either?[44]

More interesting in this respect, however, appears—at least at first glance— the *encounter scenario*. Here we can convince ourselves, at least this is what a large number of science fiction films signal, by our own eyesight, 'who we are

---

[44] At this point, one could put forward the thesis that the confrontation with a 'machine civilization' would rather increase human anxiety because of the lack of biological commonalities.

dealing with'—biological or artificial beings.[45] But how sure are we here? Since we don't know what an alien species' bodies look like, we can't tell them apart from their artificial emissaries (who, after all, may still have been created in their image). And the AI-controlled representatives of a post-biological secondary civilization could be physically developed (here also: built) in such a way that we think they are organic beings, especially since we have no idea what their original creators once looked like. This will probably only be clarified at the moment when our understanding has advanced so far (if it ever gets there) that even relatively abstract questions can be asked: Were you born or built? (We assume here that the questions actually asked will be somewhat more elaborate). At this point, however, we can venture a thesis: If we should come into direct contact with a hybrid civilization, there is much to suggest that its emissaries are more likely to come from the artificial than organic 'branch' of the civilization in question.[46] For biological entities, whose species has evolved on the protected surface of a planet or moon, the journey through the vastness of hostile space represents a significantly greater challenge than for artificial beings, who may even have been created for the sole purpose of surviving the millennia-long journey to an alien planetary system as unharmed as possible (Shostak 2015, p. 950).

Ultimately, however, we are likely to be confronted here with the problem of *general indistinguishability*: are the alien entities that visit us on Earth biological beings protected by highly advanced technologies (such as protective suits or artificial body armor), are they artificial emissaries of a biological primary or hybrid civilization, or are we confronted with the representatives of a postbiological secondary civilization? We think that this is a question that could remain unanswered for a long time. However, and this is the crucial point in this chapter, we cannot begin an exosociological analysis of the alien civilization in earnest until this question is clearly settled. Before then, all considerations regarding the

---

[45] The *camouflage thesis*, according to which extraterrestrials create artificial bodies that look confusingly similar to human bodies after prolonged observation of the Earth for the purpose of exploring our planet or also for establishing contact as carriers of their consciousness, is excluded here (Schetsche 2008, pp. 244–245). This is a popular starting scenario in science fiction, especially in times of general fear of infiltration and invasion—see, for example, the US television series *The Invaders* from 1967.

[46] Following one of the possible resolutions of the Fermi paradox, namely the consideration that post-biological secondary civilizations have little interest in contact with a biological species and wait accordingly with the establishment of contact until 'the machines' have also taken over on Earth, one could also conclude: If direct contact takes place, we are probably dealing with a biological or with a hybrid civilization. However, this assumes that the premise of these considerations is correct. In this respect, too, we are anything but certain.

structure of the alien civilization, as well as regarding its motives and interests, remain even more speculative than they are likely to be at first contact anyway.

References

Berger, Peter L. & Thomas Luckmann. 1966. *The Social Construction of Reality. A Treatise in the Sociology of Knowledge*. Garden City, NY: Doubleday.

Bohlmann, Ulrike M. & Moritz J. F. Bürger. 2018. Anthromorphism in the Search for Extra-Terrestric Intelligence – The Limits of Cognition? *Acta Astronautica* 143: 163–168.

Bostrom, Nick. 2014. *Superintelligenz. Szenarien einer kommenden Revolution*. Frankfurt am Main: Suhrkamp.

Bostrom, Nick. 2017. *Superintelligence. Paths, Dangers, Strategies* (Reprinted with corrections). Oxford: Oxford University Press.

Chick, Garry. 2014. Biocultural Prerequisites for the Development of Interstellar Communication. In *Archaeology, Anthropology, and Interstellar Communication*, Ed. Douglas A. Vakoch, 203–226. Washington: NASA.

Cleland, Carol E. & Christopher F. Chyba. 2002. Defining ‚Life'. *Origins of Life and Evolution of the Biosphere* 32: 387–393.

Crutzen, Paul J. 2002. Geology of Mankind. *Nature* 415: 23.

Crutzen, Paul J., Mike Davis, Michael D. Mastrandrea, Stephen H. Schneider, & Peter Sloterdijk. 2011. *Das Raumschiff Erde hat keinen Notausgang. Energie und Politik im Anthropozän*. Berlin: Suhrkamp.

Dick, Steven J. 2003. Cultural Evolution, the Postbiological Universe, and SETI. *International Journal of Astrobiology* 2 (1): 65–74.

Ellis, Erle C. 2018. *Anthropocence. A Very Short Introduction*. Oxford: Oxford University Press.

Fischer, Joachim. 2009. *Philosophische Anthropologie. Eine Denkrichtung des 20. Jahrhunderts*. Freiburg im Breisgau: Karl Alber.

Gehlen, Arnold. 1986. *Der Mensch. Seine Natur und seine Stellung in der Welt*. Wiesbaden: Aula-Verlag.

Giesen, Bernhard. 2010. *Zwischenlagen. Das Außerordentliche als Grund der sozialen Wirklichkeit*. Weilerswist: Vellbrück.

Grunwald, Martin. 2012. Das Sinnessystem Haut und sein Beitrag zur Körper-Grenzerfahrung. In *Körperkontakt. Interdisziplinäre Erkundungen*, Hrsg. Renate-Berenike Schmidt und Michael Schetsche, 29–54. Gießen: Psychosozial-Verlag.

Harrison, Albert A. 1993. Thinking Intelligently about Extraterrestrial Intelligence: An Application of Living Systems Theory. *Behavioral Science* 38 (3): 189–217.

Herzing, Denise L. 2014. Profiling Nonhuman Intelligence: An Exercise in Developing Unbiased Tools for Describing other „Types" of Intelligence on Earth. *Acta Astronautica* 94: 676–680.

Hölldobler, Bert & Edward Wilson. 2010. *Der Superorganismus. Der Erfolg von Ameisen, Bienen, Wespen und Termiten*. Berlin, Heidelberg: Springer.

Kutschera, Ulrich. (4. Aufl.) 2015. *Evolutionsbiologie*. Stuttgart: UTB.

Lestel, Dominique. 2014. Ethology, Ethnology, and Communication with Extraterrestrial Intelligence. In *Archaeology, Anthropology, and Interstellar Communication*, Hrsg. Douglas A. Vakoch, 227–234. Washington: NASA.

Levin, Samuel R., Thomas W. Scott, Helen S. Cooper & Stuart A. West. 2017. Darwin's Aliens. *International Journal of Astrobiology.* https://doi.org/10.1017/S1473550417000362. Zugegriffen: 15. Januar 2018

Martinez, Claudio L. Flores. 2014. SETI in the Light of Cosmic Convergent Evolution. *Acta Astronautica* 104: 341–349.

Morris, Simon Conway. 2003. The Navigation of Biological Hyperspace. *International Journal of Astrobiology* 2 (2): 149–152.

Raybeck, Douglas. 2014a. Predator-Prey Models and Contact Considerations. In *Extraterrestrial Altruism. Evolution and Ethics in the Cosmos*, Ed. Douglas A. Vakoch, 49–63. Berlin, Heidelberg: Springer.

Raybeck, Douglas. 2014b. Contact Considerations a Cross-Cultural Perspective. In *Archaeology, Anthropology, and Interstellar Communication*, Ed. Douglas A. Vakoch, 142–158. Washington: NASA.

Scheler, Max. (Orig. 1928) 2016. *Die Stellung des Menschen im Kosmos*. Berlin: Contumax.

Schetsche, Michael. 2008. Auge in Auge mit dem maximal Fremden. Kontaktszenarien aus soziologischer Sicht. In *Von Menschen und Außerirdischen. Transterrestrische Begegnungen im Spiegel der Kulturwissenschaft*, Hrsg. Michael Schetsche und Martin Engelbrecht, 227–253. Bielefeld: transcript.

Schetsche, Michael, René Gründer, Gerhard Mayer & Ina Schmied-Knittel. 2009. Der maximal Fremde. Überlegungen zu einer transhumanen Handlungstheorie. *Berliner Journal für Soziologie* 19 (3): 469–491.

Schetsche, Michael & Ina Schmied-Knittel. 2018. Zur Einleitung: Heterodoxien in der Moderne. In *Heterodoxie. Konzepte, Traditionen, Figuren der Abweichung*, Hrsg. Michael Schetsche und Ina Schmied-Knittel, 9–33. Köln: Herbert von Halem.

Schulze-Makuch, Dirk & William Bains. 2017. *The Cosmic Zoo. Complex Life on Many Worlds*. Cham: Springer Nature.

Shostak, Seth. 2015. Searching for Clever Life. *Astrobiology* 15 (11): 948–950.

Smart, John M. 2012. The Transcension Hypothesis: Sufficiently Advanced Civilizations in Variably Leave our Universe and Implications for METI and SETI. *Acta Astronautica* 78: 55–68.

Smith, John Maynard & Eörs Szathmáry. 1995. *The Major Transitions in Evolution*. Oxford, New York: Freeman and Company.

Stevenson, David S. & Sean Large. 2017. Evolutionary Exobiology: Towards the Qualitative Assessment of Biological Potential on Exoplanets. *International Journal of Astrobiology.* https://doi.org/10.1017/S1473550417000349. Zugegriffen: 18. Januar 2018

Storch, Volker, Ulrich Welsch & Michael Wink. (3. Aufl.) 2013. *Evolutionsbiologie*. Heidelberg: Springer.

Toepfer, Georg. 2011. Leben. In *Historisches Wörterbuch der Biologie. Geschichte und Theorie der biologischen Grundbegriffe*, Bd. 2. Stuttgart: Metzler: 420–483.

Traphagen, John W. 2014. Culture and Communication with Extraterrestrial Intelligence. In *Archaeology, Anthropology, and Interstellar Communication*, Hrsg. Douglas A. Vakoch, 159–172. Washington: NASA.

Urban, Tim. 2015. The AI Revolution: Our Immortality or Extinction. https://waitbutwhy.com/2015/01/artificial-intelligence-revolution-1.html and https://waitbutwhy.com/2015/01/artificial-intelligence-revolution-2.html.

Vakoch, Douglas A. 2014. The Evolution of Extraterrestrials. The Evolutionary Synthesis and Estimates of the Prevalence of Intelligence beyond Earth. In *Archaeology, Anthropology, and Interstellar Communication*, Ed. Douglas A. Vakoch, 189–202. Washington: NASA.

Wason, Paul K. 2014. Inferring Intelligence. Prehistoric and Extraterrestrial. In *Archaeology, Anthropology, and Interstellar Communication*, Ed. Douglas A. Vakoch, 112–128. Washington: NASA.

Zaun, Harald. (Hrsg.) 2010. *Kosmologie – Intelligenzen im All*. Hannover: Heise.

Hot Potato in Scientific Alien Research

11

The *ancient aliens hypothesis,* the *UFO phenomenon* as well as *alien abductions*—three very special subject areas that have been causing problems for scientific alien research for decades. They are, as one would say in everyday language, *hot potatoes* of any preoccupation with extraterrestrials. There are two closely related reasons for this precarious status: First, they are sets of questions that have been dominated by *lay research* for decades.[1] And secondly, all three topics have been *ridiculed* by the mass media ever since there has been a public discourse about them, not consistently, but very frequently.[2] The consequence of both is that scientists and scholars who are concerned with safeguarding their professional reputation (and almost all of them are) only deal with these topics in emergencies; for most of them the danger of being exposed to ridicule in the professional community and in public seems too great. Why these subject areas seem so 'ridiculous', discursively downright 'impossible', is one of the two questions to be addressed in this chapter. The other is devoted to the insights that can be gained by changing the scientific perspective when considering the three aforementioned thematic complexes.

Fortunately, unlike others, we need have no fear of contact at this point. This is due to the fact that *exosociology,* when asking about social thinking about extraterrestrials (and the related processes of tabooing), can orientate itself on the unofficial guiding principle of the sociology of knowledge: Every cultural body of knowledge, without exception, is worthy of investigation! Under this premise, it is

---

[1] On the problem of lay research, especially in scientific frontier areas, cf. Schetsche (2004).

[2] A discourse strategy widely used in the media public sphere to culturally delegitimize (heterodox) bodies of knowledge perceived as deviating from the mainstream (Berger and Luckmann 1991, p. 123; Mayer 2003; Schetsche 2013, pp. 398–399; Schetsche 2015, pp. 65–66).

legitimate from a social-scientific point of view to take a closer look at precisely those topics that pose a problem for scientific research on extraterrestrials. Before we ask about the epistemological value of such a perspective for exosociology, we would like to introduce the three 'hot potatoes' mentioned above a little more closely.

11.1   Ancient Aliens Hypothesis

The core of the ancient aliens hypothesis consists in the idea that extraterrestrials visited the earth in the prehistoric or early times of mankind, that there was contact between humans and extraterrestrials, which influenced human development culturally or even biologically—and that the extraterrestrials would have left detectable traces of their presence on earth. The ancient aliens hypothesis gained worldwide notoriety through Swiss author Erich von Däniken, whose works have sold millions of copies worldwide. In 1968, Däniken published his first book, *Erinnerungen an die Zukunft* (English title: *Chariots of the Gods?*), which made him instantly famous. In it, Däniken claimed that the history of the development of human civilization could only be understood at all by assuming an extraterrestrial influence. However, the extraterrestrials had not only influenced mankind, but had even created modern man:

> The gods of the dim past have left countless traces which we can read and decipher today for the first time because the problem of space travel, so topical today, was not a problem, but a reality, to the men of thousands of years ago. I claim that our forefathers received visits from the universe in the remote past, even though I do not yet know who these extraterrestrial intelligences were or from which planet they came. I nevertheless proclaim that these 'strangers' annihilated part of mankind existing at the time and produced a new, perhaps the first, *homo sapiens*. (von Däniken 1973, p. VIII; emphasis in original)

The book initially started with a print run of 6,000 copies. Four years later, 1.3 million copies had already been sold. According to Döring-Manteuffel, Däniken's success can also be explained by contemporary history:

> Däniken took advantage of the race into space that had been going on for years between the bloc powers, culminating in 1969 with the sensational moon landing by the Americans. This event aroused fears but also curiosity among the population. In the following years, what Däniken had recognized early on was confirmed. What was at stake was fundamental: the future of humanity, utopian designs for the renewal

of the world, the destruction of nature, and the fear of overpowering technology. (Döring-Manteuffel 2008, p. 223)

For Däniken it is certain that the extraterrestrial visitors must have appeared to humans like 'gods' due to their technological superiority, therefore various god myths are to be taken more or less literally and would in truth describe the contact of mankind with extraterrestrials. As 'circumstantial evidence' for his claims Däniken considers various archaeological finds and buildings from prehistoric and historic times, which for him can only be explained under the assumption of an influence of extraterrestrial visitors. After Däniken had made the ancient aliens hypothesis known all over the world, a multitude of mostly lay scientific authors took up the topic and dealt with the alleged traces of former extraterrestrial visitors in numerous publications. In these works there exists a kind of *canon* of alleged evidence for the visit of extraterrestrials, which, in addition to passages from various mythical traditions and sacred writings, includes in particular various archaeological artifacts, some of which are indeed astounding, which, according to the representatives of the ancient aliens hypothesis, do not fit into the established scientific picture of the history of human civilization. A well-known example of this is a stone relief from Palenque, a richly decorated stone tomb cover featured on the cover of *Chariots of the Gods?* The stone slab covered the sarcophagus of the Maya ruler K'inich Janaab Pakal, who ruled Palenque in the seventh century. In Mayan archaeology, the images on the stone slab are interpreted as representing Pakal's journey to the underworld (Xibalbá). Erich von Däniken and other proponents of the the ancient aliens hypothesis, on the other hand, see it as a depiction of an "astronaut" in a "rocket". Däniken writes (Fig. 11.1).

A genuinely unprejudiced look at this picture would make even the most die-hard skeptic stop and think. There sits a human being, with the upper part of his body bent forward like a racing motorcyclist; today any child would identify his vehicle as a rocket. It is pointed at the front, then changes to strangely grooved indentations like inlet ports, widens out, and terminates at the tail in a darting flame. The crouching being himself is manipulating a number of indefinable controls and has the heel of his left foot on a kind of pedal. His clothing is appropriate: short trousers with a broad belt, a jacket with a modern Japanese opening at the neck, and closely fitting bands at arms and legs. With our knowledge of similar pictures, we should be surprised if the complicated headgear were missing. And there it is with the usual indentations and tubes, and something like antennae on top. Our space traveller – he is dearly depicted as one—is not only bent forward tensely; he is also looking intently at an apparatus hanging in front of his face. The astronaut's front seat is separated by struts from the rear portion of the vehicle, in which symmetrically arranged boxes, circles, points, and spirals can be seen. (von Däniken 1973, pp. 98–99)

The "unbiased", "naive" view that Däniken describes here is sometimes elevated to a "method" within ancient aliens hypothesis. Archaeological artifacts and historical writings are *not to be* interpreted within the framework of scientifically recognised methods, but are to be subjected to what Däniken calls a "contemporary view". Specifically, this means that various archaeological traces must be interpreted against the background of modern developments in technology—and only in this way can they be properly understood. In other words:

Archaeological as well as textual indications are not considered by the paleo-SETI in their cultural context, but in the context of the hypothetical global paleo-contact. The central frame of reference is modern (including futuristic) conceptions of technology; the pre-astronautical interpretation of the circumstantial evidence is almost always an

Fig. 11.1 Book cover Erich von Däniken: "Erinnerungen an die Zukunft" [Englisch title: "Chariots of the Gods?"]. *Source* Scan of the book cover

interpretatio technologica [...] in which alien concepts are equated with concepts from one's own horizon of understanding. (Richter 2015, p. 353; emphasis in original).

Erich von Däniken's theses were rejected from the beginning by the scientific community due to a lack of evidence and unscientific methods, but this did not diminish the commercial success of his books. Quite the opposite: For 40 years now, von Däniken has successfully staged himself as a taboo-breaking outsider who pushes mainstream science before him, questions outdated dogmas and thus acts as a scientific revolutionary. However, Däniken does not seriously engage with scientific criticism of his theses, but essentially ignores it. There is nothing to add to the conclusion of Richter (2015, p. 355) in this context: "In my opinion, paleo-SETI research is currently not using the available scope for scientific dialogue by demanding recognition, insisting on its position."

With all justified and necessary criticism of the approach of Erich von Däniken and other representatives of the ancient aliens hypothesis, we would like to emphasize, however, that we consider the question of whether there could be legacies of extraterrestrial visitors on Earth to be legitimate *in principle*—but its answer must be guided by scientific standards. Erich von Däniken and his successors are undoubtedly to thank for the fact that the ancient aliens hypothesis still enjoys great popularity, but in the end they have probably done a disservice to the *scientific* investigation of possible extraterrestrial artifacts on Earth. The ancient aliens hypothesis has an aura of unseriousness that makes serious scientific efforts in this direction enormously difficult.

11.2 The UFO Phenomenon

A broad spectrum of human experiences is subsumed under the keyword 'UFO phenomenon', ranging from simple sightings of light phenomena in the sky that appear unusual to complex experiences with unknown beings and objects. The term 'UFO' (short for unidentified flying object) originated in a military context and was initially a term for all flying objects that could not be clearly classified due to their characteristics. This is what it says in a dossier of the *US Air Force* from 1954:

Unidentified flying objects (UFOB)—Relays to any airborne object which by performance, aerodynamic characteristics, or unusual features does not conform to any presently known aircraft or missile type, or which cannot be positively identified as a familiar object. (Air Force Regulation 200-2)

Independently of this, reports of strange 'flying objects' or 'airships' have appeared again and again in recent centuries, which have always puzzled people, but at least since the historic sighting of the US businessman and amateur pilot Kenneth Arnold in 1947 have also been interpreted as *'flying saucers'* in the sense of extraterrestrial spaceships. Soon those flying saucers were also called 'UFOs' and since then the association of the term 'UFO' with the interpretation that at least some of the sightings could be extraterrestrial spaceships has become firmly established. A significant contribution to this has been made by a wide variety of fictional references to the subject in cinema films, TV series, comics and novels, in which UFOs are almost invariably portrayed as extraterrestrial spaceships. The term 'UFO' now refers to a firmly anchored (and largely globalised) pattern of interpretation that is applied whenever a specific observation, individually or collectively, does not seem to fit into the framework of conventional explanatory schemes. Hardly a day goes by without an observation being made in the sky or on the ground somewhere in the world that cannot initially or even permanently be interpreted by the eyewitnesses with the help of conventional explanatory models and is thus referred to as a 'UFO' (Schmied-Knittel and Wunder 2008, p. 133) At least some of these sightings are ultimately also interpreted by the witnesses in terms of extraterrestrial visitors. This immediately reveals the analytical fuzziness of the common use of the term, which, as we had already pointed out elsewhere (Schetsche and Anton 2013, p. 10), contains at least three different levels of meaning. The term 'UFO' today means:

I In everyday language, anything unusual in the sky that cannot be classified by observers in any other way—for example, as an aircraft (this is ultimately a residual category denoting the abnormal);
II in the original aeronautical-military sense, any object (or, more generally, phenomenon) in the Earth's atmosphere or near-Earth space not currently identified by the relevant experts (such as pilots or radar observers), and
III the interpretation of a sighting in terms of (I) or (II) as an object or phenomenon produced by an extraterrestrial civilization (this is usually referred to as the 'ET hypothesis').

We emphasize this because these levels often get confused in communication about the UFO phenomenon. If, in everyday life, an individual observer undifferentiated refers to a celestial phenomenon that he or she cannot classify otherwise as a 'UFO', this should be regarded as a potentially *hybrid* classification that can refer both to the status of the provisionally unexplained in the sense of 'UFO I' *and* to a sometimes more, sometimes less seriously meant interpretive hypothesis in the sense of 'UFO III'. The consequences of foregoing a corresponding terminological differentiation are particularly evident in the mass media's portrayals of the topic: when, for example, witnesses speak about their perceptions in terms of 'UFO I' or 'UFO II', this is regularly portrayed by the mass media as if the people concerned had made an interpretation in terms of 'UFO III'— an extraneous attribution that is regularly used to ridicule the people concerned and delegitimise their testimony (and thus the observation itself) (Schetsche and Anton 2013, p. 11).

Despite some efforts, research into the UFO phenomenon has not yet made it into the canon of accepted science. In our opinion, one of the reasons for this is that the term 'UFO' is usually understood and rejected by critics of UFO research in the sense of the third level of meaning. The idea of the impossibility of bridging interstellar distances plays a special role in this context (for example: Drake and Sovbel 1994, p.102).

With the justified reference to the difficulty of interstellar spaceflight, unjustified criticism or discredit is often levelled at UFO research as a whole (Michaud 2007, p. 153; Wendt and Duvall 2012, p. 285). Unjustified because the extraterrestrial hypothesis is by no means the focus for scientific UFO research and those UFO researchers who consider the extraterrestrial hypothesis as an explanation for unexplained sightings are likely to be in the minority (Hövelmann 2008, p. 169). So this is, as it were, a straw man argument. Even more: the assumption of the existence of extraterrestrial civilisations is in no way constitutive for UFO research, in contrast to the SETI programme. *Interpreting* UFOs as extraterrestrial spacecraft represents one of several possibilities discussed within UFO research. In accordance with the scientific guiding principle of strictly separating observations and interpretations, the scientifically oriented definitions of the term 'UFO' do not contain any references to extraterrestrial spaceships as a possible explanation for the phenomenon. For example, astronomer Hynek's (1972, p. 26) definition, which remains one of the most common determinations of the term 'UFO', states:

We can define the UFO simply as the reported perception of an object or light seen in the sky or upon the land, the appearance, trajectory, and general dynamic and luminescent behavior of which do not suggest a logical, conventional explanation and which is not only mystifying to the original percipients but remains unidentified after close scrutiny of all available evidence by persons who are technically capable of making a common sense identification, if one is possible.

Even in the various classification systems for mysterious celestial phenomena developed within UFO research (Anton and Ammon 2015, p. 333), there are no references to extraterrestrial spaceships, although these are repeatedly brought into play as a *possible explanatory hypothesis for* unexplained UFO sightings. Regardless of this, we consider the reference to the alleged unbridgeability of interstellar distances to be fundamentally problematic, since it contains anthropocentric presuppositions at its core and argues against the background of our *current* state of technical knowledge what will be possible or even impossible in the future. Moreover, the rejection of the extraterrestrial hypothesis in the context of the UFO phenomenon, insofar as it comes from representatives of the SETI paradigm, also seems to be *strategically* motivated.[3] Since its inception, the SETI program, fully committed to the remote contact scenario, has struggled to secure long-term funding. It is clear that if other contact scenarios were also promising and perhaps even more cost-effective to implement, a competitive situation would immediately arise in terms of public attention and economic resources. The suspicion suggests itself that the technical feasibility and sufficient financial viability of the search, but not paradigmatic and theoretical meaningfulness, were and are in the foreground here, entirely in the sense of Drake's programmatic statement: "Let's just put up receivers" (Sheridan 2009, p. 67). UFO researchers, on the other hand, are faced with a completely different starting situation: their research efforts are still much less scientifically recognised than those of SETI researchers; the financing of large-scale research programmes is currently not even conceivable—for the entire UFO research of the last decades, not even a fraction of the sums of money were available that a single current SETI project claims today (Pirschl and Schetsche 2013, p. 43).[4]

Despite the difficulties inherent in the UFO phenomenon as an object of study, there have been repeated attempts in the past to investigate it using classical *scientific methods.* If one summarises the few academic studies that have explicitly dealt with the UFO topic, a very inconsistent picture emerges. What is certain is that the vast majority of sightings can be explained by stimuli that have been misinterpreted by the witnesses, such as bright stars, planets, comets, satellites, aircraft or insects, lens flares and light reflections in photographs, etc., can be explained. However, experience shows that there is a certain percentage (about 5 percent) of UFO sighting cases that cannot be given a clear conventional explanation even after thorough investigations. Some researchers see in this—and

---

[3] A systematic sociological analysis of this relationship has already been presented by Romesberg (1992).

[4] On the 'strategic ignorance' of science towards UFO research, see Dodd (2018).

Fig. 11.2   UFO over the Black Forest—artistic representation. *Source* original graphic for this volume by Nadine Heintz

especially in individual spectacular sighting cases—indications of the existence of a hitherto not understood phenomenon (or rather: several phenomena), others assume that these cases could also be explained conventionally with sufficient data (Fig. 11.2).

In our view, the current state of knowledge regarding the UFO phenomenon in no way justifies the activities of some UFO enthusiasts, some of which can be described as missionary, who consider the 'proof' of extraterrestrial visitors to have been established long ago, and thinkslittle of the sweeping criticism and discrediting of scientific research into the UFO phenomenon. Decisive gains in knowledge about the UFO phenomenon could be expected in the context of pro-fessionalised academic-scientific UFO research, but such research appears almost unthinkable in the foreseeable future due to the epistemological, methodological-methodological and, above all, the special status of the UFO phenomenon in terms of science policy (Anton and Ammon 2015, p. 344).[5]

---

[5] We would like to make an update at this point as an exception, because the situation has fundamentally changed since 2018 (when we wrote the German version of the book). From the end of 2017, various information about the UFO phenomenon became public in the U.S.,

11.3 Alien Abductions

Here we are confronted with what are probably the most bizarre theses regarding extraterrestrials from the point of view of our scientific and public dominant culture when we deal with the most complex of topics that in English-speaking countries are called *alien abduction experiences*.

The reports of experiences, which we are dealing with here, originate—apart from a few precursors—from the sixties to the nineties of the last century.[6] In these decades, people from all over the world—their total number is difficult to estimate, but may have amounted to several hundred thousand—report that they were abducted by extraterrestrial beings for hours or days in their spaceships, where gruesome medical experiments were performed on them. The aliens are described as technologically superior, yet frighteningly ruthless and unsympathetic.[7]

The narratives of the self-declared abduction victims vary in details, but they agree in an astonishing way especially in the passages that can be called structural (so Whitmore 1993; Newman and Baumeister 1996, pp. 100–102; Showalter 1997, p. 258; Brookesmith 1998, pp. 7–9; Bullard 2003, p. 88). In particular, the sequence of experiences is almost always described in a similar way, so that a homogeneous basic pattern of abductions emerges—one of us (Schetsche 2008, p. 158) had characterized this *abduction pattern in* an earlier text thus:

1. The victim first sees an unusual celestial phenomenon or awakens to a brilliantly bright light.

---

which had a massive impact on public communication about the subject. In the meantime, two official UAP reports are available and UFOs were the subject of a hearing in the US Congress. In view of this development, one can rightly speak of a paradigm shift: UFOs are considered a serious topic in the U.S., a potential threat to national security, and are to be systematically investigated. This 'destigmatization' means that UFOs are increasingly becoming a legitimate academic research topic, which we expressly welcome—and which we would hardly have thought possible until a few years ago (cf. Anton & Vugrin 2022).

[6] A sociologically interesting phenomenon that remains unexplained today is the sharp decline in these abduction accounts at the turn of the millennium. In the twenty-first century, the abduction narrative seems culturally outdated in more ways than one. (Which does not mean that the authors of this volume do not occasionally still receive corresponding reports from German-speaking countries as well).

[7] An overview of the phenomenon and public debates about it is provided by Bynum (1993), Spanos et al. (1993), Newman and Baumeister (1996), Paley (1997), Schetsche (1997), Lynn et al. (1998).

2. strange figures appear out of nowhere, robbing the affected person of willpower and sensation by unknown methods.
3. through these figures (or through some kind of light beam) the victim is taken to a brightly lit room, often filled with alien machines, which is supposed to be on board a spaceship.
4. here—usually fixed on some kind of table or bed—it is subjected to various, often very painful examinations or experiments: blood and tissue samples are taken, thin probes are inserted into various body orifices or through the skin, sometimes implants are inserted.
5. In many cases, the abductors are particularly interested in the reproductive apparatus of the abductees. Sperm or eggs are taken, in some cases there is sexual interaction between humans and human-like aliens.
6. at the end of the investigations, either the memories of the events are erased or the victims' minds are manipulated so that they cannot talk about their experiences."

It is striking that the basic tenor of the abduction reports has changed noticeably since the early variants in the 1950 and 1960s. Thus, many victims and observers initially assumed rather positive after-effects of the abduction experience: Interaction with the aliens was said to have helped the abductees gain 'higher insights' about themselves or the future of humanity. In the overwhelming majority of later reports (from the 1970s onwards), however, the abductions and their after-effects are then judged *extraordinarily negatively*. In particular, the medical experiments performed on the victims are interpreted as highly *traumatizing* experiences; it is said that the majority of the abductees suffered from symptoms of post-traumatic stress disorder after their return (Vacarr 1993; Whitmore 1993, pp. 316–317; Newman and Baumeister 1996, p. 100; Porter 1996; Goldberg 2000, p. 311; Johnson 1994; Cromie 2003).

Since the end of the 1980s, there has been a whole series of scientific publications on the subject in the English-speaking world. The authors—in contrast to the victims—regularly assume that the alien abductions *did not really happen* in the way described. The scientific literature is therefore dominated from the beginning by a *critical* view of the *phenomena*. The reports of experiences are mostly interpreted psychologically-psychiatrically, for example as a consequence of a basic psychological disorder of the self-declared abduction victims or also as 'cover memories', with which a real traumatizing event (for example a sexual assault in childhood) is to be hidden from the own psyche.[8] This attempt

---

[8] For the details of the scholarly debate, see Schetsche (2008, pp. 159–164).

at an explanation may be plausible from a psychological point of view when interpreting some individual cases, but it does not explain the astounding similarities in thousands of abduction reports. This is probably also the reason why the explanatory hypothesis arose in the literature in the 1990s that the victims had all become victims of the *False Memory Syndrome* in a similar way (Schnabel 1994; Spanos et al. 1994; Lynn and Kirsch 1996; Newman and Baumeister 1996; Orne et al. 1996; Lynn et al. 1998). This thesis is more plausible than the classical psychopathological explanations in that the emergence of alien abduction memories is seen here as the result of a therapeutic process in which widespread *social interpretive patterns* play a central role. A culturally influenced iatrogenic reality construction is assumed,[9] at the end of which those affected are subjectively certain that they have been abducted by aliens and subjected to terrible experiments (Fig. 11.3).

Following these reflections, one of us (Schetsche 2008) had proposed an *integrative model* for explaining the phenomenon, in which psychological factors are complemented by cultural ones. According to this understanding, alien abductions represent a *psychosocial phenomenon* that is well known to the public thanks to extensive reporting in the mass media, but whose reality status is highly disputed between scientific experts, lay researchers and those affected. From the point of view of the sociology of knowledge, this phenomenon belongs equally to two segments of reality[10]:

> As a narrative of human confrontation with nonhuman, extraterrestrial actors, it comes from the world of the fictional-fantastic. As subjectively certain memories of such confrontations, it belongs to the collectively shared reality of those affected, their supporters, and the audience inclined to believe in the reality of those accounts. The social reality of abductions is thereby constituted in a feedback loop between public thematization and individual victim careers; in it, published accounts of victims and the forms of psychotherapeutic practice – which both evoke and are guided by them – play the decisive role [...]. Against the background of corresponding media

---

[9] The method of *regression hypnosis*, to which many victims owe their (supposed) memories, is of particular importance here (according to Newman and Baumeister 1996, p. 105; McLeod et al. 1996, p. 16). This method had gained acceptance among therapists in the 1970s and 1980s in the context of treating victims of sexual or ritual abuse, especially in the USA, but eventually came under criticism because experimental and clinical studies showed that actual memories obtained under hypnosis cannot be reliably distinguished from pseudo-memories produced in the hypnotic process (see Streeck-Fischer et al. 2001, p. 20; Fiedler 2001, p. 116; Schacter 2001, p. 443).

[10] As an exception, we quote here in somewhat greater detail from an earlier text, since the state of scientific knowledge concerning the phenomenon has hardly developed further in the last ten years.

Fig. 11.3 Typical scene from the abduction narrative—artistic representation. *Source* original graphic for this volume by Nadine Heintz

reporting, unspecific symptoms feed the suspicion in some recipients that they themselves belong to the group of sufferers. A closer examination of the subject, such as attending lectures or reading non-fiction books and autobiographies on the subject, creates the need to find out for sure, which drives those concerned into the arms of self-help groups, lay hypnotists or 'open-minded' therapists. In a continuous conversation with other 'victims' or under hypnosis, the suspicion finally becomes a supposed certainty: in the interactive process, more and more images appear that can be put together to form an abduction scenario. Fuzzy fragments of memory become detailed images and scenes, unspecific symptoms become specific signs with the status of evidence. Piece by piece, a subjectively certain knowledge about a traumatizing experience, the abduction by aliens, emerges. At the end of this process, the subject is left with the unshakable certainty of having been the victim of aliens in the past. An initially medial reality has thus been transformed into a subjective reality, which is, however, to a large extent an intersubjectively shared one due to the social origin of the interpretive logic and the symbolic material used. The abduction reports are thus not structurally similar because they would be based on matching contents of memories or even identical real experiences, but because unspecific symptoms, short memory fragments and bizarre dream images are interpreted and transformed into individual memories with the help of collectively shared knowledge stocks and interpretation logics. In this process, the media representations provide the basic narrative and visual material for the production of memories, and the published memories in turn provide the basis for new media representations. (Schetsche 2008, pp. 170–171).

This model is able to explain scientifically the vast majority of abduction reports without having to assume the action of extraterrestrial powers on Earth (which would be an extremely serious presupposition in more ways than one). What remains, however, are questions concerning that minority of cases in which the memories could not have arisen in the context of a therapeutic process, perhaps because the experiences were never forgotten. The greatest problem in explanation, however, is posed by abduction accounts that date back to before the interpretive pattern presented was culturally common. It is striking that many of these cases on this side of mass media reporting differ markedly in their sequence of events from the later narratives (Bullard 1999, p. 188).

Regardless of which scientific interpretation of the extraordinary reports of experiences may have the greatest explanatory power, these nocturnal 'encounters' at least allow us to deduce something about the *ideal* relationship between humans and extraterrestrials (Schetsche 2008, p. 174). We will return to this at the end of this chapter.

11.4 Exosociological Conclusions

What insights can be gained by considering the three aforementioned thematic fields from the perspective of the sociology of knowledge and then attempting to make this perspective fruitful for a future exosociology? We begin with three brief answers regarding the individual topic areas and then, at the very end of this chapter, venture a brief programmatic synopsis.

1. **Ancient aliens hypothesis**: With the theses of the then so-called "pre-astronautics", which became popular especially through the works of Erich von Däniken, the idea has spread globally that mankind has received one or more visits of extraterrestrial intelligences in the distant past. If one classifies this conviction not, as is usually done,[11] in terms of religious studies, but in terms of the sociology of knowledge, one can observe here the emergence and spread of a *heterodox body of knowledge* whose reality content is vehemently denied by the mainstream of (historical) science, but which nevertheless, or precisely because of this, enjoys great public popularity. From the point of view of the sociology of knowledge, it is not decisive whether there is sufficient historical evidence for von Däniken's basic thesis, but rather that the idea of an earlier cultural contact, indeed of a spiritual fertilization—or even

---

[11] See, for example, the extremely knowledgeable volume by Richter (2017).

control—of the first advanced human cultures by extraterrestrial visitors has become *conceivable* worldwide through the work of the Swiss lay researcher. The cultural innovative power of ancient aliens hypothesis lies, in our view, precisely not in the creation of a religion-like edifice of thought, but on the contrary in the *secularization of* creation myths and religious worldviews: The emissaries of technological civilizations take the place of transcendent gods as *supernatural* agents of human history with the power to act. Even if the earthly 'evidence' for this early cultural contact cannot stand up to the critical eyes of archaeologists and historians, the current state of knowledge of astrophysics and astrobiology today leaves little doubt that, at least theoretically, such visits by extraterrestrials to Earth *could* have occurred (we have discussed this in detail in various chapters of this volume[12]). Whether this must have happened in historical or at least mythologically traditional times is an entirely different question. In the end, the program, carried out by lay researchers, provides less convincing circumstantial evidence for such a visit than stimulating building blocks for thinking—primarily in life-world terms—about man's place in the cosmos. With regard to this subtopic of exosociology, the debates surrounding the ancient aliens hypothesis are therefore well worth a more intensive look.

2. **UFO sightings**: Although similar phenomena existed long before, the UFO theme as we know and understand it today is nevertheless a cultural product of the twentieth century. In terms of discourse history, it developed at the intersection of nation-state action (military policy, intelligence disinformation programmes), swelling popular enthusiasm for space in East and West, and the mass media's struggle for sales figures and ratings. In this discursive space, the actual observations (i.e. the subjective or occasionally also intersubjective evidentiary experiences of unusual celestial phenomena) regularly receded into the background. Instead of asking phenomenologically what exactly the UFO-witnesses actually saw, their reports are mostly ignored by the scientific community—and the eyewitnesses themselves are often pathologized. The reason for this is the *discourse-strategic* 'misunderstanding' that still dominates in the mass media as well as in the scientific public, which equates the sighting of an "unidentified flying object" by often rather *helpless* eyewitnesses with an interpretation of those optical stimuli as 'extraterrestrial spaceships' manned by 'little green men'.[13] This is an interpretation that is

---

[12] Thus the *necrological scenario* we examine in Chap. 8 assumes such contact in prehistoric times.

[13] A detailed explanation of this shift in meaning and its consequences is provided by Hövelmann (2008), Anton (2013) and Schetsche and Anton (2013).

regularly applied to such sightings from the outside and then stuck on them, as it were, like a *faulty label*. The subjective and intersubjective evidence of UFO observations undoubtedly belongs to the field of research of scientific anomalistics[14] (we therefore want to bracket this category of questions here), the mass-media and scientifically imposed 'ET interpretation', on the other hand, makes the phenomenon exosociologically interesting: via the erroneous interpretation of the UFO concept, the topic of the visit of extraterrestrials to Earth has been kept culturally virulent since the middle of the last century, but at the same time epistemologically defused through fictionalisation. Thus, the mostly mocking media coverage of unusual celestial phenomena meets with fictional representations (novels, films, TV series, computer games) of the alien theme and produces a melange of postmodern thinking about an alien presence in the immediate vicinity of the earth that is difficult to separate analytically. In the context of the UFO theme, an alien visit to Earth at least seems *conceivable, if not necessarily a serious possibility*. And the aliens themselves are constituted as celestial observers who, for whatever reason, may not be officially imagined by us humans. The perpetual mysteriousness of their motives and interests is what makes almost all corresponding fictional representations and probably also many serious reports so appealing.

3. **Abductions by extraterrestrials**: A completely different picture of 'extraterrestrial visitors' is painted by the *abduction narrative* closely associated with the UFO theme. Here, for decades,[15] it has been dominated by the image of ruthlessly vicious aliens who abduct people against their will (or in a state of unconsciousness) into their spaceships in order to carry out medical experiments on them. At least some of these experiments, which in their cold objectivity coupled with absolute ruthlessness are ultimately reminiscent of the human experiments carried out by German concentration camp doctors in the Third Reich, seem to have *reproductive medical* goals. The corresponding subtype of the abduction reports leads us into a world of breeding a new race of humans or even the creation of human-alien hybrids. From a cultural-historical perspective, we find ourselves here at the heart of the great narrative of *transhumanism*—at least this part of the abduction narratives emphatically points beyond the thought traditions of classical modernity. From an exosociological point of view, the abduction narratives are so interesting (quite

---

[14] See the article "UFO Sichtungen" (Anton and Ammon 2015, pp. 332–345) in the anthology *An den Grenzen der Erkenntnis*, ed. Mayer et al.

[15] The 'contactee' variant that dominated in the fifties and sixties, reporting friendly invitations to tour the solar system aboard alien spaceships, must—we had already pointed out—be regarded as discursively outdated.

independently of their reality status) because they manifest the basic *xeno-phobic* mood that has always 'gone along' with thinking about extraterrestrials in cultural history. The reports of self-declared abduction victims, as well as media coverage (rare) or fictionalization (more frequent), generate a *sense of threat* in the cultural consciousness that is, admittedly, cryptic, but for that very reason constantly present. It nourishes the question of what else can be expected from 'the aliens' in view of their infinite technical superiority when they have one day completed their breeding experiments, which *are inhumane in both senses. Will* they then move on without a word or will they continue to carry out their plans, which are completely incomprehensible to us (but presumably highly perfidious)? In the event of a first contact (culturally recognized in its reality!) between humanity and an extraterrestrial civilization, the xenophobic abduction narratives could be a factor, not necessarily rational but collectively effective, in assessing the dangerousness of an extraterrestrial species. Ruthless experimenters who play up their technological superiority without moral scruples are not something anyone wants as cosmic neighbors.

Let us turn to some final **implications for exosociology**:

All three topics should be kept in view exosociologically. In particular, it should be clarified what significance they each have for the development of cultural thinking about extraterrestrials; one of the most important keywords here is *xenophobia*. Accordingly, it should also be asked what role these bodies of knowledge might have in assessing the consequences of a real first contact. (In our scenario analyses we had largely excluded these three special complexes of topics in order not to further increase the complexity of our forecasts).

Furthermore, in order not to get tangled up in the pitfalls of a reductionism of cultural studies (which only allows for the epistemological perspective), one should definitely also ask the question about the *potential reality content of* the corresponding theses, experiences, narratives. In all three subject areas, we can and should at least occasionally also investigate *phenomena*:

- How likely are prehistoric and early historic alien visits to Earth? What evidence of their presence could they have left behind in principle? Why are those 'evidences', which lay research persistently presents to us, so unsuitable from a scientific point of view? But which indications for an earlier presence of extraterrestrials on Earth would we scientifically accept at all?
- What would be, if at least a small part of the so far unexplained celestial phenomena (more than five percent of the sighting reports are not at issue here anyway) were actually extraterrestrial spacecraft? What insights could

be derived from the sightings and the occasional technical data beyond the certainty now attained that the Fermi paradox would have to be considered obsolete? And which options of an 'interplanetary' communication would still be open to us as long as the visitors refuse a systematic exchange of information?

- How would we as humans deal with it politically and morally if aliens abducted a few thousand 'of us' every year and subjected them to painful experiments? What, for instance, would change for our earthly self-image if we knew that an extraterrestrial power willing to experiment could degrade us to helpless victims at will? What options would be available at all to put an end to these assaults? Or should we as a global society tolerate the terrible fate of individuals without complaint in order to avoid the risk of a military confrontation with a technically probably far superior opponent?

So as not to be misunderstood at this point: We personally take a very critical view of the reality content of all three narratives. But this does not change the fact that it is one of the tasks of exosociology not only to keep an eye on the cultural discourses on these topics, but also to ask what 'could be true' on an ontological or phenomenological level to the respective theses, experiences, reports. Exosociology should not expose itself to the reproach of unquestionable *ontological gerrymandering*.[16] In a modified form, the dictum of the *wild card concept* might apply here: Even extremely improbable assumptions are worthy of closer examination if the possible consequences of their correctness are serious enough.

References

Air Force Regulation 200–2 or AFR 200–2. Version August 1954. https://en.wikisource.org/wiki/Air_Force_Regulation_200–2_Unidentified_Flying_Objects_Reporting.
Anton, Andreas. 2013. Zur (Un-)Möglichkeit wissenschaftlicher UFO-Forschung. In *Diesseits der Denkverbote. Bausteine für eine reflexive UFO-Forschung*, Hrsg. Michael Schetsche und Andreas Anton, 49–77. Hamburg: Lit.

---

[16] An important term in the sociology of social problems, with which structural-functionalist problem theory has reacted to the constructionist criticism of its basic assumptions concerning the reality content of social problems—from our point of view, by and large for good reasons: It is scientifically inadmissible to claim the irrelevance of social facts to the emergence and development of public discourse unless one is prepared to subject the question of relevance to *empirical* scrutiny, that is, to relate social facts themselves to debates about those facts (on this, see Woolgar and Pawluch 1985 and Schetsche 2000, pp. 18–23).

Anton, Andreas & Danny Ammon. 2015. UFO-Sichtungen. In *An den Grenzen der Erkenntnis. Handbuch der wissenschaftlichen Anomalistik*, Hrsg. Gerhard Mayer, Michael Schetsche, Ina Schmied-Knittel und Dieter Vaitl, 332–345. Stuttgart: Schattauer.

Anton, Andreas & Fabian Vugrin. 2022. "UFOs exist and everyone needs to adjust to that fact." (Dis)Information Campaigns on the UFO Phenomenon. *Journal of Anomalistics* 22: 18–35. https://doi.org/10.23793/zfa.2022.18

Berger, Peter L. & Thomas Luckmann. (Deutsche Erstausgabe 1969) 1991. *Die gesellschaftliche Konstruktion der Wirklichkeit. Eine Theorie der Wissenssoziologie.* Frankfurt am Main: Fischer.

Brookesmith, Peter. 1998. *Alien Abductions.* London: Blandford.

Bullard, Thomas E. 1999. What's New in UFO Abductions? Has the Story Changed in 30 Years? *MUFON Symposium Proceedings* 1999: 170–199.

Bullard, Thomas E. 2003. False Memories and UFO Abductions. *Journal of UFO Studies* 8: 85–160.

Bynum, Joyce. 1993. Kidnapped by an Alien. Tales of UFO Abductions. *ETC. A Review of General Semantics* 50: 86–95.

Cromie, William J. 2003. Alien Abduction Claims Examined: Signs of Trauma Found. *Harvard University Gazette*. https://news.harvard.edu/gazette/story/2003/02/alien-abduction-claims-examined-2/.

Däniken, Erich von. 1968. *Erinnerungen an die Zukunft.* Düsseldorf, Wien: Econ.

Däniken, Erich von. 1973. *Chariots of the Gods? Unsolved Mysteries of the Past.* News York: Bantam Books.

Dodd, Adam. 2018. Strategic Ignorance and the Search for Extraterrestrial Intelligence: Critiquing the Discursive Segregation of UFOs from Scientific Inquiry. *Astropolitics. The International Journal of Space Politics and Policy* 16 (1): 75–95.

Döring-Manteuffel, Sabine. 2008. *Das Okkulte. Eine Erfolgsgeschichte im Schatten der Aufklärung. Von Gutenberg bis zum World Wide Web.* München: Siedler.

Drake, Frank & Dava Sobel. 1994. *Signale von anderen Welten. Die wissenschaftliche Suche nach außerirdischer Intelligenz.* München: Droemer.

Fiedler, Peter. 2001. *Dissoziative Störungen und Konversion. Trauma und Traumabehandlung.* Beinheim: Beltz/PVU.

Goldberg, Carl. 2000. The General's Abduction by Aliens from a UFO: Levels of Meaning of Alien Abduction Reports. *Journal of Contemporary Psychotherapy* 30: 307–320.

Hövelmann, Gerd. 2008. Vernünftiges Reden und technische Rationalität. Erkenntnistheoretische Überlegungen zu Grundfragen der UFO-Forschung. In *Von Menschen und Außerirdischen. Transterrestrische Begegnungen im Spiegel der Kulturwissenschaft*, Hrsg. Michael Schetsche und Martin Engelbrecht, 183–204. Bielefeld: transcript.

Hynek, J. Allen. 1972. *The UFO Experience. A Scientific Inquiry.* Chicago: Henry Regnery.

Hynek, J. Allen. 1979. *UFO. Begegnungen der ersten, zweiten und dritten Art.* München: Goldmann.

Johnson, Ronald C. 1994. Parallels between Recollections of Repressed Childhood Sex Abuse, Kidnappings by Space Aliens, and the 1692 Salem Witch Hunts. *Issues in Child Abuse Accusations* 6 (1): 41–47.

Lynn, Steven Jay, und Irving I. Kirsch. 1996. Alleged Alien Abductions: False Memory, Hypnosis, and Fantasy Proneness. *Psychological Inquiry* 7 (2): 151–155.

Lynn, Steven Jay, Judith Pintar, Jane Stafford, Lisa Marmelstein & Timothy Lock. 1998. Ren-
dering the Implausible Plausible: Narrative Construction, Suggestion, and Memory. In
Believed-in Imaginings: The Narrative Construction of Reality, Eds. Joseph de Rivera
and Theodore R. Sarbin, 123–143. Washington: American Psychological Association.
Mayer, Gerhard. 2003. Über Grenzen schreiben. Presseberichterstattung zu Themen aus
dem Bereich der Anomalistik und der Grenzgebiete der Psychologie in den Printmedien
SPIEGEL, BILD und BILD AM SONNTAG. *Zeitschrift für Anomalistik* 3: 8–46.
McLeod, Caroline C., Barbara Corbisier & John E. Mack. 1996. A More Parsimonious
Explanation for UFO Abduction. *Psychological Inquiry* 7 (2): 156–168.
Michaud, Michael A. G. 2007. *Contact with Alien Civilizations. Our Hopes and Fears about
Encountering Extraterrestrials*. New York: Springer.
Newman, Leonard S. & Roy F. Baumeister. 1996. Toward an Explanation of UFO Abduc-
tion Phenomenon: Hypnotic Elaboration, Extraterrestrial Sadomasochism, and Spurious
Memories. *Psychological Inquiry* 7 (2): 99–126.
Orne, Martin M., Wayne G. Whitehouse, Emily Carota Orne & David F. Dinges. 1996.
‚Memories of Anomalous and Traumatic Autobiographical Experiences: Validation and
Consolidation of Fantasy through Hypnosis. *Psychological Inquiry* 7 (2): 168–172.
Paley, John. 1997. Satanist Abuse and Alien abduction: A Comparative Analysis Theorizing
Temporal Lobe Activity as a Possible Connection between Anomalous Memories. *The
British Journal of Social Work* 27: 43–70.
Pirschl, Julia & Michael Schetsche. 2013. Aus Fehlern lernen. Anthropozentrische Voran-
nahmen im SETI-Paradigma – Folgerungen für die UFO-Forschung. In *Diesseits der
Denkverbote. Bausteine für eine reflexive UFO-Forschung*, Hrsg. Michael Schetsche und
Andreas Anton, 29–48. Berlin: Lit.
Porter, Jennifer E. 1996. Spiritualists, Aliens and UFOs: Extraterrestrials as Spirit Guides.
Journal of Contemporary Religion 11: 337–353.
Richter, Jonas. 2015. Paläo-SETI. In *An den Grenzen der Erkenntnis. Handbuch der wis-
senschaftlichen Anomalistik*, Hrsg. Gerhard Mayer, Michael Schetsche, Ina Schmied-
Knittel, und Dieter Vaitl, 346–347. Stuttgart: Schattauer.
Richter, Jonas. 2017. *Götter-Astronauten. Erich von Däniken und die Paläo-SETI-Mythologie*.
Hamburg: Lit.
Romesberg, Daniel Ray. 1992. *The Scientific Search for Extraterrestrial Intelligence: A
Sociological Analysis*. Ann Arbor: UMI Dissertation Services.
Schacter, Daniel L. 2001. *Wir sind Erinnerung. Gedächtnis und Persönlichkeit*. Reinbek bei
Hamburg: Rowohlt.
Schetsche, Michael. 1997. „Entführungen durch Außerirdische" – ein ganz irdisches Deu-
tungsmuster. *Soziale Wirklichkeit* 1: 259–277.
Schetsche, Michael. 2000. *Wissenssoziologie sozialer Probleme. Grundlegung einer relativis-
tischen Problemtheorie*. Wiesbaden: Westdeutscher Verlag.
Schetsche, Michael. 2004. Zur Problematik der Laienforschung. *Zeitschrift für Anomalistik*
4: 258–263.
Schetsche, Michael. 2008. Entführt! Von irdischen Opfern und außerirdischen Tätern. In *Von
Menschen und Außerirdischen. Transterrestrische Begegnungen im Spiegel der Kulturwissenschaft*, Hrsg. Michael Schetsche und Martin Engelbrecht, 157–182. Bielefeld:
transcript.

Schetsche, Michael. 2013. Unerwünschte Wirklichkeit. Individuelle Erfahrung und gesellschaftlicher Umgang mit dem Para-Normalen heute. *Zeitschrift für Historische Anthropologie* 21: 387–402.

Schetsche, Michael. 2015. Anomalien im medialen Diskurs. In *An den Grenzen der Erkenntnis. Handbuch der wissenschaftlichen Anomalistik*, Hrsg. Gerhard Mayer, Michael Schetsche, Ina Schmied-Knittel und Dieter Vaitl, 63–73. Stuttgart: Schattauer.

Schetsche, Michael & Andreas Anton. 2013. Einleitung: Diesseits der Denkverbote. In *Diesseits der Denkverbote. Bausteine für eine reflexive UFO-Forschung*, Hrsg. Michael Schetsche und Andreas Anton, 7–27. Hamburg: Lit.

Schmied-Knittel, Ina & Michael Schetsche. 2003. Psi-Report Deutschland. Eine repräsentative Bevölkerungsumfrage zu außergewöhnlichen Erfahrungen. In *Alltägliche Wunder. Erfahrungen mit dem Übersinnlichen – wissenschaftliche Befunde*, Hrsg. Eberhard Bauer und Michael Schetsche, 13–38. Würzburg: Ergon.

Schmied-Knittel, Ina & Edgar Wunder. 2008. UFO-Sichtungen. Ein Versuch der Erklärung äußerst menschlicher Erfahrungen. In *Von Menschen und Außerirdischen. Transterrestrische Begegnungen im Spiegel der Kulturwissenschaft*, Hrsg. Michael Schetsche und Martin Engelbrecht, 133–155. Bielefeld: transcript.

Schnabel, Jim. 1994. Chronicles of Aliens Abduction and some other Traumas as Self-Victimization Syndrom. *Dissociation: Progress in the Dissociative Disorders* 7 (1): 51–62.

Sheridan, Mark A. 2009. *SETI's Scope: How the Search for ExtraTerrestrial Intelligence Became Disconnected from New Ideas about Extraterrestrials.* Ann Arbor: ProQuest.

Showalter, Elaine. 1997. *Hystorien. Hysterische Epidemien im Zeitalter der Medien.* Berlin: Berlin Verlag.

Spanos, Nicholas P., Patricia A. Cross, Kirby Dickson & Susan C. DuBreuil. 1993. Close Encounters: An Examination of UFO Experiences. *Journal of Abnormal Psychology* 102: 624–632

Spanos, Nicholas P., Cheryl A. Burgess & Melissa Faith. 1994. Past-life Identity, UFO Abductions, and Satanic Ritual Abuse: The Social Construction of Memories. *The International Journal of Clinical and Experimental Hypnosis* XLII (4): 433–446

Streeck-Fischer, Annette, Ulrich Sachsse & Ibrahim Özkan. 2001. Perspektiven in der Traumaforschung. In *Körper, Seele, Trauma*, Hrsg. Annette Streeck-Fischer, Ulrich Sachsse und Ibrahim Özkan, 12–22. Göttingen: Vandenhoeck & Ruprecht

Vacarr, Barbara Adina. 1993. *The Divine Container. A Transpersonal Approach in the Treatment of Repressed Abduction Trauma.* Ph.D., The Union Institute (Cincinnati/Ohio). Demand Copy: University Microfilms International.

Wendt, Alexander & Raymond Duvall. 2012. Militanter Agnostizismus und das UFO-Tabu. In *Generäle, Piloten und Regierungsvertreter brechen ihr Schweigen*, Hrsg. Leslie Kean, 281–294. Rottenburg: Kopp.

Whitmore, John. 1993. Religious Dimensions of the UFO Abductee Experience. *Syzygy: Journal of Alternative Religion and Culture* 2 (3–4): 313–326.

Woolgar, Steve & Dorothee Pawluch. 1985. Ontological Gerrymandering: The Anatomy of Social Problems Explanations. *Social Problems* 32: 214–227.

Outlook

12

A major scientific work usually ends with some kind of outlook. This is not easy to do here, because our book itself is in part *futurologically* oriented, and the concept of the outlook is therefore inherent in it. Nevertheless, we try to do so and in this concluding chapter we first look back at the contents of our book in the form of a short conclusion, then look forward to the future of exosociology, and finally look upwards, figuratively speaking, towards the sky and its possible inhabitants.

12.1 Looking Back

In what is now considered his classic excursus on 'the stranger', the German sociologist Georg Simmel wrote in 1908:

> The inhabitants of Sirius are not actually alien to us—this at least not in the sociologically considered sense of the word – *but they do not exist for us at all,* they stand beyond far and near. (Simmel 1958, p. 509; emphasis by the authors).

This quotation shows only too clearly: even great thinkers are intellectual children of their time. In 1908, aliens existed for sociology only in the form of *imagined* figures in early utopian or dystopian narratives (*The War of the Worlds* by H. G. Wells had appeared ten years earlier) or—even more unpleasantly for the just-establishing scientific discipline—as thought experiments of a cosmologically oriented metaphysics. One hundred and ten years later, *now*, it is time to bid farewell to Simmel's verdict. As we had shown, the extraterrestrials have left the fictional world and have become *real in* a sociological sense, namely phenomenologically—for the time being still as a hypothetical counterpart, but nevertheless with the claim to face us in the flesh one day in the distant (but perhaps also near) future as *maximum strangers* (Schetsche et al. 2009). Before this happens, first contact is only an imaginable 'disruptive event' in the context of the scientific view of the future (Steinmüller and Steinmüller 2004, pp. 86–87). After the event has occurred, however, the strangers become actors with almost unlimited agency. This is the reason why sociology—unlike in Simmel's time—can no longer, we are convinced, refuse to take note of the aliens as *real factors* (in Scheler's sense—1924, pp. 5–6) and just take them seriously accordingly. This is exactly what we have tried to do in our book (the following section can also be read as a *topic overview*).

In the introduction (Chap1) we explained why *now* is the time to resume the program of exosociology after decades and to deal intensively with *extraterrestrial intelligences from a* social science perspective: Sociology should, we firmly believe, respond to the accumulating evidence from scientific research that the existence of extraterrestrial life seems more likely than ever before in human history. We explained that the new hyphenated sociology will rest methodologically and theoretically on two pillars: *futurology* and alien *studies*. We also discussed the current goals of exosociology, including those three thematic complexes that were given particular importance in the volume presented: the critical monitoring of SETI research, the prognosis of the consequences of humanity's encounter with an extraterrestrial intelligence, and the possibilities and limitations of the sociological analysis of extraterrestrial civilizations.

Chapter 2 has provided a brief overview of the idea of extraterrestrials in *Western thought*. From the philosophical speculations of antiquity, the path went through the Renaissance, which with the heliocentric worldview provided us with groundbreaking insights into the structure of our solar system, to the threshold of cultural modernity, in which an intense interplay between fictional and scientific thought unfolded. This interplay has ultimately remained with us to this day when pondering those 'aliens'. It is difficult to decide analytically whether it was science fiction that provided the imaginations and cues for scientific thinking about extraterrestrials, or whether the scientific findings (on the habitability of distant planets, for example) fueled the imaginations of writers and filmmakers more strongly. When we look at those images we form of the "maximum alien" today, we are confronted with an inseparable melange of tradition-rich philosophical thought, diverse scientific findings, and fictional representations of the most varied kinds. This finding is of sociological significance whenever it is a question of understanding today's alien images not only of the public, but also of the scientific community. This concerns the scientific search for extraterrestrials as well as the prognosis of the possible consequences of an interstellar first contact.

Chapter 3 explained why this contact is to be expected in the near or at least more distant future. Using the famous 'Drake formula', it was explained how likely it is to encounter life outside the Earth and possibly also—certainly highly alien—intelligences. The chapter provides, in the end probably rather untypical for a sociological volume, an overview of the *scientific state of knowledge on* the subject complex 'extraterrestrial life and extraterrestrial intelligence'. The comprehensiveness of our presentation resulted on the one hand from our conviction that the further argumentation in the volume would not be comprehensible if some basic information on the position of the Earth in the cosmos and on the probability of extraterrestrial life were not first provided—on the other hand, however, from our (in the meantime empirically saturated) experience that this knowledge cannot be assumed without further ado by many sociological colleagues.

Building on this basic knowledge, Chap. 4 is devoted to the now almost 70-year history of the radio astronomical search for extraterrestrial signals. Under the acronym *SETI*, which has gained a certain popularity through media coverage, a comparatively small number of enthusiasts (and an even smaller number of female enthusiasts) try to pick up messages from alien intelligences from the vastness of space. The fact that this project has not been successful up to the present day can simply be regarded as 'bad luck'—or as a consequence of questionable presuppositions which, at least according to our thesis in that chapter, result from anthropocentric prejudices that could have been avoided. Here, the decades-long isolation of SETI research contoured by natural science from philosophical, linguistic and social science knowledge takes its revenge. Even if this self-blinding has weakened somewhat in recent years, the review of more than one hundred SETI projects of the recent past must nevertheless be rather critical from a sociological point of view.

The consequences that could be drawn from the failure of conventional SETI research were demonstrated in Chap. 5 using an alternative exploration strategy—the *search for extraterrestrial artifacts* in our solar system (SETA). This research direction is still young, and has so far (not least for financial reasons) tended to present theoretical concepts rather than practical experiments or even large-scale exploration projects. However, it is gaining more scientific legitimacy with each successive year that no radio message has been received from extraterrestrial intelligences. The aim of this chapter was to reconstruct the research logic of this 'other search programme', which is still largely unknown to the public, and to comment on it from an anthropological and sociological point of view.

A completely different consequence of the very practical failure of traditional SETI research is drawn by those scientists (who also mostly come from the field of radio astronomy) who are trying to switch from passive listening to an active transmission mode. In Chap. 6 we have presented the basic idea of those *METI projects,* which have recently caused at least occasional public furore. The media attention results not least from the fact that these experiments are highly controversial in the scientific community and many experts (such as the recently deceased astrophysicist Stephen Hawking) have emphatically warned of the incalculable consequences. From the point of view of the sociology of science, as should be clear from reading this chapter, the METI projects represent *high-risk research* in the existential sense. The various project plans in this area have reached a stage in which internal scientific criticism is no longer sufficient, but rather a normative intervention by state and multinational institutions seems necessary.

Chapter 7 goes one step further, in which we explained why the prognostic assessment of the *consequences* of humanity's contact with an extraterrestrial intelligence is one of the main tasks of exosociology. The chapter takes as its starting point the futurological concept of so-called *wild cards*; these are events that are very rare but, when they do occur, have massive and lasting effects on terrestrial civilization as a whole. *Scenario analysis* introduced the futurological method by which the consequences of such events can be predicted. Here we favor the form of narrative analysis, which provides dense descriptions of the expected effects of first contact.

The methodological considerations were put into practice in Chap. 8: Here we presented the results of the *scenario analysis* we had carried out. In each case, first abstractly and then exemplarily-concretely, we predicted the consequences of different variants of first contact between mankind and an extraterrestrial civilization. The *signal scenario* (we receive a technically generated radio signal from the far reaches of our galaxy), the *artifact scenario* (we find an undoubtedly extraterrestrial artifact in the asteroid belt), and the *encounter scenario* (an alien spacecraft enters the Earth's orbit under control) were examined. In the context of the *necrology scenario* (the discovery of an extraterrestrial mummy in the permafrost soil), we also discussed the possibility that knowledge of an actual first contact in prehistoric times would not find social acceptance, and that the corresponding convictions would therefore remain culturally marginalized. The scenarios we have worked out circumscribe a technical-structural space in which contacts with extraterrestrials could unfold if the worst came to the worst.

Following the scenario analysis, Chap. 9 dealt with the question of whether *systematic preparation* of humankind for first contact is already possible and sensible today. We first discussed the general problems of such preparation (such as today's prevailing anthropocentrism or the widespread political ignorance of the problem) and then presented some current proposals from the scientific debate that could remedy this situation. We presented our own ideas on the topic in the form of *four concrete strategies* for minimizing the possible negative effects of first contact (here, among other things, we dealt with the basic features of crisis communication). The chapter concluded with five *guiding principles* that should be taken into account by scientific, governmental and international institutions when preparing for first contact.

From a sociological point of view, Chap. 10 is certainly another centerpiece of the book. Under the heading "Proto-Sociology of Extraterrestrial Civilizations", the question was asked what can be said today about extraterrestrial civilizations on the basis of earthly presuppositions and theories—and what cannot be said. The chapter started from what we consider to be the analytically central differentiation between *biological* and *post-biological* societies. While little can be said about the latter type today, *evolutionary theory* allows for a number of inferences regarding the development, structure, and functioning of civilizations dominated by biological species. We have presented here a structured set of questions that might guide future analysis of such societies. Further subchapters dealt with the role of space-related *technologies* of alien civilizations as well as with the special case of a *hybrid civilization* in which biological species and artificial intelligences cooperate.

Finally, in Chap. 11, we dealt with three issues that remain 'hot potatoes' for extraterrestrial scientific research to this day: The *ancient aliens hypothesis* claims that humanity received visitations from extraterrestrial intelligences in its prehistory or early history. The *UFO phenomenon* deals with celestial phenomena that cannot be readily explained and are therefore often interpreted as extraterrestrial spaceships. And the stories of *abductions by extraterrestrials* confront us with subjective evidential experiences that are difficult to accept scientifically. All three topics are directly related to the question of the existence of extraterrestrial intelligences, but are located beyond traditional (natural) scientific research and pose considerable challenges to it. We have re-analyzed the aforementioned topics from the *perspective of the sociology of knowledge* and thus reclassified them scientifically. This certainly does not solve all scientific problems of access to these topics, but at least enables them to be treated within the framework of exosociology without the otherwise culturally customary disparagement.

12.2   Looking Ahead

In the introduction we had already indicated that in the sub-discipline of sociology that we have brought back into play, a distinction must be made between two phases with regard to its tasks and its scientific *and* social relevance: There is an exosociology I *before* first contact and there will be—this can be said prognostically—an exosociology II *after* first contact. Both variants start with similar initial questions, but differ with regard to their main tasks. The contact of mankind with an extraterrestrial civilization will also mean a *caesura* for exosociology.

In Phase I, the new sub-discipline should, according to our ideas, concentrate on five sets of questions—we had already introduced these in the introduction, so a repetition of the 'headings' will suffice here. The task of exosociology thereafter is:

1. the critical sociological accompaniment of the scientific search for extraterrestrials;
2. the prognostic estimation of the terrestrial consequences of a future first contact;
3. the study of the interrelations between scientific and fictional reflection on the place of man in the cosmos;
4. the contouring of a research on foreignness and xenophobia that focuses on non-human actors;
5. participation in the social discussions on the main features of an extra-human ethics.

After mankind has gained certain knowledge of the existence of extraterrestrial intelligences, however, exosociology—now having entered Phase II—will probably have to concern itself primarily with three tasks:

(a) the analysis of the structures and functioning, developments and goals of that extraterrestrial civilization of which we have become aware;
(b) the *empirical* study of the concrete consequences of first contact for human societies and for terrestrial civilization as a whole;
(c) the critical monitoring of the various phases of cultural contact between the terrestrial and extraterrestrial civilization(s).

ad (a) We had already discussed in detail in Chap. 10 (proto-sociology) how we imagine this task from today's point of view—i.e. *before the first contact*. We do not need to repeat that here. It should be clear that from the moment of this contact (however it will be concretely shaped) almost everything will be different—for mankind, for science in general and, of course, also for exosociology. On the basis of the *empirical* data that will then be collected, we must once again carefully consider which of our prospectively formulated questions can actually be posed to the alien civilization and which are excluded for certain reasons. How extensive exosociological studies *can* subsequently turn out to be depends primarily on whether the knowledge we have acquired about the alien intelligence is rather meagre or whether we have richly bubbling sources of information at our disposal. In this respect, the first contact scenarios we have discussed (Chap. 8) differ very fundamentally. The question of what the alien intelligence allows us to learn about them is also likely to play a role. In this respect, already the manner of their *information control* (which we expect) generates some knowledge about the alien civilization. Since all this is still in the future, little can be said today about this first point beyond these general indications. What is clear, on the other hand, is that at the moment of first contact exosociology will change from a futurological to an *empirical-phenomenological* science with an object of study situated in the present. At that moment, it will be important to have our sociological tools of data collection and data analysis at hand and to adapt them promptly to the study of a concrete alien civilization. How well this succeeds will become apparent in the event.[1]

ad (b) Before the initial contact, all knowledge about the various *social consequences of* the event is subject to great *uncertainty*. We had discussed this in the abstract in Chap. 7 (Methodological Considerations), but also repeatedly referred to this central problem point in Chap. 8 (Scenario Analysis). What we can say about the possible consequences of first contact from today's perspective and using classical futurological methods is necessarily limited—much *must* remain uncertain. From the moment of first contact, however, these estimates will acquire a real data basis: The various human societies and terrestrial civilization as a whole will react to the news of first contact with an extraterrestrial intelligence. We will then be able, on the basis of what will certainly be rapidly accumulating empirical material, to make statements about the factual reactions of social groups and different cultures, about the influence of religions and scientific debates, about nation-state and transnational action, and so on. This is admittedly not a task for which sociology alone would be responsible; rather, other disciplines (such as psychology, political science, religious studies, and economics)

---

[1] We think: The instruments of exosociology are for mankind in the universe something like the 'first aid kit' for the earthly car: One hopes never to need it—but if one does, one is glad to have it.

must join in the research and have their say here. But at least sociology, with the considerations presented at that time by Jan H. Mejer and now also by us, has a concept to present, within the framework of which and with the help of which the upcoming global processes of change can be taken into consideration, scientifically reconstructed and also socially and ethically evaluated. In the current state of scientific reflection on the consequences of first contact, sociology should therefore play the role of a *leading discipline in* the event (at least that is our hope[2]). The event is likely to be accompanied by significant changes in the research landscape in general, which will not spare the social and cultural sciences. We are already seeing chairs for 'interstellar communication' and 'global transformation' springing up—at least if the contact proceeds in a way that gives the human science and education system as we know it today any chance of survival. This seems to us anything but certain, at least in the case of *direct contact*; it is possible that the extraterrestrials will bring with them methods of education and research on their visit to Earth which are based on completely different principles and which we humans will be only too willing to adapt. Based on Earthly experience, we know that there is considerable *cultural pressure to adapt from* civilizations that appear technologically superior (we had discussed this in Chap. 8). Perhaps, however, human life on Earth will develop under alien influence in such a way that there will be no need at all for future-oriented research and education. In that case, the 'Phase II' of exosociology that we predict will be correspondingly short.

ad (c) We have well-founded doubts as to whether exosociology will play a significant role in the very practical shaping of communication[3] with an extraterrestrial intelligence. We are realistic enough to know: The days of sociology as the leading scientific discipline of modern societies have been over for decades. When it comes to first contact, other disciplines will certainly come to the fore— and they will claim with greater vehemence (and higher chances of success) their jurisdiction over 'the aliens' and everything to do with them: Astrobiology, Linguistics, Computer Science, Political Science, Philosophy and probably

---

[2] A role that is likely to be disputed by (social) psychology, which has also been dealing with the question of first contact for decades (see the groundbreaking work of Albert Harrison). Of course, we also advocate interdisciplinary cooperation at this point, as we already do today in the "Research Network Extraterrestrial Intelligence" (www.eti-research.net)—but in terms of professional policy, it makes perfect sense to claim a leading role for sociology at this point, at least for a moment.

[3] In this case, only remote contact (where the formulation of response messages is involved) and direct contact (with, hypothetically, a variety of communication channels) come into question anyway.

even Theology. Which discipline then gains the upper hand in political discourse will depend in particular on how the contact process is shaped, how much the communication processes can be monopolized, and who holds the reins on the human side: military and intelligence agencies, the governments of nation states, international organizations, or perhaps multinational corporations. If the political structures known to us (in the broadest sense) survive the first contact phase at all[4], at some point the hour will surely come for *scientific experts* to advise this or that actor with the power to act, perhaps even to make one or two tactical decisions. But that is all. In terms of their organisational forms and their modes of knowledge production, the sciences are not intended as decision-making bodies and cannot relieve political, military and economic actors of their responsibility. Even if some SETI researchers may already see themselves as 'great communicators' and 'hypercultural mediators' between us and the aliens,[5] we should nevertheless remain realistic at this point and content ourselves with an advisory role. Sociology, as a tried and tested instance of social self-observation (Luhmann 1986), has the *additional* task of keeping an eye on the counselling processes themselves—and critically questioning its own role in the process. In this context, we console ourselves with the fact that there is always something relieving about not having to decide on the weal and woe of humanity. And at least in an encounter scenario, 'the others' will probably make the most important decisions anyway.

12.3 The (Sociological) View Upwards

Who has come to this point in the book and knows a little about the SETI literature, will have noticed that our explanations differ significantly at two points from what has been discussed among the 'alien researchers' coming from the natural sciences in the last decades: First, we are clearly more skeptical about the communication possibilities with the extraterrestrials. Here we tend to follow the assessment of Russian scientists such as Sukhotin (1971) and therefore do not think that it will be readily possible to decode signals received from space or even symbols on an extraterrestrial artefact. It may be that this will take a very long

---

[4] We will not tire of emphasizing this for good reasons, but in a meeting scenario this depends first and foremost (and in a figurative sense possibly at the same time last and foremost) on how friendly and peaceful the meeting is.

[5] We assume that this tendency to overestimate oneself has to do with the erroneous equation of claims to rationality and political agency. Science fiction with an affinity for science is only too happy to serve this deceptive self-image (as in the film *Arrival*, USA 2016).

time, perhaps even fail completely. Therefore, in the case of contact in the form of such scenarios, it is conceivable that we will acquire very little information about the alien intelligence. And second, we are much more pessimistic than the mainstream of SETI researchers about the *consequences of* this first contact. We do not think that a technically highly developed civilization is automatically so far developed ethically that it will put its own interests on the back burner in case of doubt in order not to harm an alien species. Altruism and utilitarianism are *human* concepts that may be completely meaningless to an alien intelligence. But even if that is not the case, terrestrial history shows very powerfully how great the difference can be between such ideals and the actual actions of nation-states and multinational corporations. We therefore also belong to the group of those researchers who strongly warn against attempts to actively establish contact (keyword: METI). But even if the extraterrestrials should be friendly to us (or at least neutrally disposed towards us), first contact—especially if it takes the form of the encounter scenario—can have repercussions for the Earth, lead to cultural upheavals that permanently shake human social orders as a whole.

All this does not mean, however, that we are driven by a clammy (or even open) joy in the analytically appearing first contact apocalypse[6]—and certainly not that our theses should become a self-fulfilling prophecy. On the contrary, we hope that our *advice for caution* (in every respect!) will lead to a prudent handling of all responsible earthly authorities, both with an alien intelligence and with the reactions of the people on our planet.

One final thought: We would love to supplement this book with a second volume—a guide for extraterrestrial intelligences on how to understand 'humanity' and how to deal with terrestrial intelligences. But that is not our task. If there are indeed other intelligences in the vastness of the universe (which we think highly probable), it may be that these civilizations too have developed institutions whose task is to observe and analyze their own forms of coexistence. And if this should be so, it may well be that some of these intelligences have produced their own form of 'exo-exosociology'—which would then, in turn, be responsible for the study of the discovered terrestrial civilization, should the case arise. Books in our sense are unlikely to be possessed by the aliens, but probably something of a functional equivalent. And then there might also be someone, or better: something, out there that will take on this task of analysis: "humanity for beginners". And if both sides have the great good fortune to be able to reciprocally receive the respective analyses one day in the distant future, perhaps they and we will

---

[6] Dystopian cinema films such as *Oblivion* (USA 2013) stage this lust for the downfall of human civilization all too gleefully.

be able to laugh heartily at the incredible 'blindness' or even 'obtuseness' of the respective other side. At this point, however, the skeptic in us rears up one last time: Can we really assume that aliens have such a thing as a sense of humor? We hope so.

References

Luhmann, Niklas. 1986. *Die Selbstbeschreibung der Gesellschaft und die Soziologie.* Gastvorlesung an der Universität Augsburg. Radiomitschnitt Bayerischer Rundfunk, 06.11.1986. https://www.youtube.com/watch?v=NIRTM3WZLqw.
Scheler, Max. 1924. *Versuche zu einer Soziologie des Wissens.* Leipzig, Berlin: Duncker & Humblot.
Schetsche, Michael, René Gründer, Gerhard Mayer, und Ina Schmied-Knittel. 2009. Der maximal Fremde. Überlegungen zu einer transhumanen Handlungstheorie. *Berliner Journal für Soziologie* 19 (3): 469–491.
Simmel, Georg. (4. Aufl.) 1958. *Soziologie. Untersuchungen über die Formen der Vergesellschaftung.* Berlin: Duncker & Humblot.
Steinmüller, Angela, und Heinz Steinmüller. 2004. *Wild Cards. Wenn das Unwahrscheinliche eintritt.* Hamburg: Murmann.
Sukhotin, Boris Viktorovich. 1971. Methods of Message Decoding. In *Extraterrestrial Civilizations. Problems of Interstellar Communication*, Hrsg. S. A. Kaplan, 133–212. Jerusalem: Keter Press.

Bibliography

Air Force Regulation 200–2 or AFR 200–2. Version August 1954. https://en.wikisource.org/wiki/Air_Force_Regulation_200-2_Unidentified_Flying_Objects_Reporting.

Akerma, Karim. 2002. *Außerirdische Einleitung in die Philosophie. Extraterrestrier im Denken von Epikur bis Hans Jonas.* Münster: Monsenstein und Vannerdat.

Almár, Iván & Paul H. Shuch. 2007. The San Marino Scale: A New Analytical Tool for Assessing Transmission Risk. *Acta Astronautica* 60 (1): 57–59.

Anders, Günter. 1970. *Der Blick vom Mond. Reflexionen über Weltraumflüge.* München: C. H. Beck.

Anton, Andreas. 2013. Zur (Un-)Möglichkeit wissenschaftlicher UFO-Forschung. In *Diesseits der Denkverbote. Bausteine für eine reflexive UFO-Forschung,* Hrsg. Michael Schetsche und Andreas Anton, 49–77. Hamburg: Lit.

Anton, Andreas & Danny Ammon. 2015. UFO-Sichtungen. In *An den Grenzen der Erkenntnis. Handbuch der wissenschaftlichen Anomalistik,* Hrsg. Gerhard Mayer, Michael Schetsche, Ina Schmied-Knittel und Dieter Vaitl, 332–345. Stuttgart: Schattauer.

Anton, Andreas & Fabian Vugrin. 2022. "UFOs exist and everyone needs to adjust to that fact." (Dis)Information Campaigns on the UFO Phenomenon. *Journal of Anomalistics* 22: 18–35. http://dx.doi.org/https://doi.org/10.23793/zfa.2022.18

Anton, Andreas & Michael Schetsche. 2014. Im Spiegelkabinett. Anthropozentrische Fallstricke beim Nachdenken über die Kommunikation mit Außerirdischen. In *Interspezies-Kommunikation. Voraussetzungen und Grenzen,* Hrsg. Michael Schetsche, 125–150. Berlin: Logos.

Anton, Andreas & Michael Schetsche. 2015. Anthropozentrische Transterrestrik. Zur Kritik naturwissenschaftlich orientierter SETI-Programme. *Zeitschrift für Anomalistik* 15: 21–46.

Arkhipov, Alexey V. 1998. Earth-Moon System as Collector of Alien Artefacts. *Journal of the British Interplanetary Society* 51: 181–184.

Ascheri, Valeria & Paolo Musso. 2002. Kosmische Missionare? In *S.E.T.I. Die Suche nach dem Außerirdischen,* Hrsg. Tobias Daniel Wabbel, 170–184. München: Beust.

Ascher, Marcia. 1991. *Ethnomathematics – A Multicultural View of Mathematical Ideas.* Pacific Grove: Brooks/Cole Publishing.

Azua-Bustos, Armando et al. 2015. Regarding Messaging to Extraterrestrial Intelligence (METI)/Active Searchers for Extraterrestrial Intelligence (Active SETI). https://setiat home.berkeley.edu/meti_statement_0.html.

Bach, Joscha. 2004. Gespräch mit einer Künstlichen Intelligenz – Voraussetzungen der Kommunikation zwischen intelligenten Systemen. In *Der maximal Fremde. Begegnungen mit dem Nichtmenschlichen und die Grenzen des Verstehens*, Hrsg. Michael Schetsche, 43–56. Würzburg: Ergon.

Ball, John A. 1973: The Zoo Hypothesis. *Icarus* 19: 347–349.

Bartholomew, Robert E. & Hillary Evansk. 2004. *Panic Attacks. Media Manipulation and Mass Delusion*. Stroud: Sutton Publishing.

Bauman, Zygmund. 2000. Vereint in Verschiedenheit. In *Trennlinien. Imagination des Fremden und Konstruktion des Eigenen*, Hrsg. Josef Berghold, Elisabeth Menasse und Klaus Ottomeyer, 35–46. Klagenfurt: Drava.

Baum, Seth D., Jacob D. Haqq-Misra & Shawn D. Domagal-Goldman. 2011. Would Contact with Extraterrestrials Benefit or Harm Humanity? A Scenario Analysis. *Acta Astronautica* 68: 2114–2129.

Baxter, Stephan & John Elliott. 2012. A SETI Metapolicy. New Directions Towards Comprehensive Policies Concerning the Detection of Extraterrestrial Intelligence. *Acta Astro nautica* 78: 31–36.

Bell, John H. 1973. The Zoo Hypothesis. *Icarus* 19: 347–349.

Benner, Stevan A., Alonso Ricardo & Matthew A. Carrigan. 2004. Is there a Common Chemical Model for Life in the Universe? *Current Opinion in Chemical Biology* 8: 672–689.

Berger, Peter L. & Thomas Luckmann. 1966a. *The Social Construction of Reality. A Treatise in the Sociology of Knowledge*. Garden City, NY: Doubleday.

Berger, Peter L. & Thomas Luckmann. (engl. Orig. 1966b) 1991. *Die gesellschaftliche Konstruktion der Wirklichkeit. Eine Theorie der Wissenssoziologie*. Frankfurt am Main: Fischer.

Berghold, Christina. 2011. *Die Szenario-Technik. Leitfaden zur strategischen Planung mit Szenarien vor dem* Hintergrund *einer dynamischen Umwelt*. Göttingen: Optimus.

Biebert, Martina F. & Michael T. Schetsche. 2016. Theorie kultureller Abjekte. Zum gesellschaftlichen Umgang mit dauerhaft unintegrierbarem Wissen. *BEHEMOTH – A Journal on Civilisation* 9 (2): 97–123.

Billingham, John. 2014. SETI: The NASA Years. In *Archeology, Anthropology and Interstellar Communication*, Ed. Douglas Vakoch, 1–21. Washington: National Aeronautics and Space Administration.

Billingham, John, et al. 1994. *Social Implications of the Detection of an Extraterrestrial Civilization*. Mountain View, CA: SETI Institute Press.

Billings, Linda. 2015. The Allure of Alien Life. Public and Media Framings of Extraterrestrial Life. In *The Impact of Discovery Life beyond Earth*, Ed. Steven J. Dick, 308–323. Cambridge: University Press.

Bitterli, Urs. 1986. *Alte Welt – neue Welt. Formen des europäisch-überseeischen Kulturkontaktes vom 15. bis zum 18. Jahrhundert*. München: Beck.

Bitterli, Urs. 1991. *Die ,Wilden' und die ,Zivilisierten': Grundzüge einer Geistes- und Kulturgeschichte der europäisch-überseeischen Begegnung*. München: Beck.

Bohlmann, Ulrike M. & Moritz J. F. Bürger. 2018. Anthropomorphism in the Search for Extra-Terrestrial Intelligence – The Limits of Cognition? *Acta Astronautica* 143: 163–168.

Bostrom, Nick. 2014. *Superintelligenz. Szenarien einer kommenden Revolution.* Berlin: Suhrkamp.

Bostrom, Nick. 2017. *Superintelligence. Paths, Dangers, Strategies* (Reprinted with corrections). Oxford: Oxford University Press.

Bourdieu, Pierre. 1993. Über einige Eigenschaften von Feldern. In *Soziologische Fragen,* Ders., 107–114. Frankfurt am Main: Suhrkamp.

Boutle, Ian A., Nathan J. Mayne, Benjamin Drummond, James Manners, Jayesh Goyal, F. Hugo Lambert, David M. Acreman & Paul D. Earnshaw. 2017. Exploring the Climate of Proxima B with the Met Office Unified Model. Astronomy & Astrophysics, March 1, 2017. https://arxiv.org/pdf/1702.08463.pdf.

Brookesmith, Peter. 1998. *Alien Abductions.* London: Blandford.

Brookings-Report. 1960. Proposed Studies on the Implications of Peaceful Space Activities for Human Affairs [A Report Prepared for the Committee on Long-Range Studies of the National Aeronautics and Space Administration by The Brookings Institution]. Washington D.C: Brookings Institution. http://www.nicap.org/papers/BrookingsCompleteRpt.pdf.

Buchter, Heike & Burkhard Straßmann. 2013. Die Unsterblichen. Eine Begegnung mit dem Technikvisionär Ray Kurzweil und den Jüngern der ,Singularity'-Bewegung. Zeit Online am 27.03.2013. http://www.zeit.de/2013/14/utopien-ray-kurzweil-singularity-bewegung.

Bullard, Thomas E. 1999. What's New in UFO Abductions? Has the Story Changed in 30 Years? *MUFON Symposium Proceedings* 1999: 170–199.

Bullard, Thomas E. 2003. False Memories and UFO Abductions. *Journal of UFO Studies* 8: 85–160.

Brosius, Hans-Bernd. 1994: Agenda-Setting nach einem Vierteljahrhundert Forschung: Methodischer und theoretischer Stillstand? *Publizistik* 39: 269–288.

Bynum, Joyce. 1993. Kidnapped by an Alien. Tales of UFO Abductions. *ETC. A Review of General Semantics* 50: 86–95.

Cantril, Hadley. 1940. *The Invasion from Mars: A Study in the Psychology of Panic.* Princeton, NJ: Princeton University Press.

Capova, Klara Anna. 2013. The Detection of Extraterrestrial Life: Are We Ready? In *Astrobiology, History, and Socienty. Life Beyond Earth and the Impact of Discovery,* Ed. Douglas A. Vakoch, 271–281. Heidelberg: Springer.

Carrigan, Richard A. Jr. 2006. Do Potential SETI Signals Need to be Decontaminated? *Acta Astronautica* 58 (2): 112–117.

Cassan, Arnaud, Daniel Kubas & Jean Philippe Beaulieu. 2012. One or More Bound Planets per Milky Way Star from Microlensing Observations. *Nature* 481: 167–169.

Chaisson, Eric J. 2015. Internalizing Null Extraterrestrial ,Signals'. In *The Impact of Discovery Life beyond Earth,* Ed. Steven J. Dick, 324–337. Cambridge: University Press.

Chick, Garry. 2014. Biocultural Prerequisites for the Development of Interstellar Communication. In *Archaeology, Anthropology, and Interstellar Communication*, Ed. Douglas A. Vakoch, 203–226. Washington: NASA

Chorost, Michael. 2016. How a Couple of Guys Built the Most Ambitious Alien Outreach Project Ever. Smithsonian.com am 26. September 2016. https://www.smithsonianmag.com/science-nature/how-couple-guys-built-most-ambitious-alien-outreach-project-ever-180960473/?no-ist.

Clarke, Arthur C. 2016. *2001: Odyssee im Weltraum – Die komplette Saga*. München: Heyne.

Cleland, Carol E. & Christopher F. Chyba. 2002. Defining ‚Life'. *Origins of Life and Evolution of the Biosphere* 32: 387–393.

Cocconi, Giuseppe & Philip Morrison. 1959a. Searching for Interstellar Communications. *Nature* 186: 670–671.

Cocconi, Giuseppe & Philip Morrison. 1959b. Searching for Interstellar Communications. *Nature* 184: 844–846.

Connolly, Bob & Robin Anderson. 1987. *First Contact*. New York: Viking Penguin.

Cromie, William J. 2003. Alien Abduction Claims Examined: Signs of Trauma Found. *Harvard University Gazette*. https://news.harvard.edu/gazette/story/2003/02/alien-abduction-claims-examined-2/.

Crutzen, Paul J. 2002. Geology of Mankind. *Nature* 415: 23.

Crutzen, Paul J., Mike Davis, Michael D. Mastrandrea, Stephen H. Schneider, & Peter Sloterdijk. 2011. *Das Raumschiff Erde hat keinen Notausgang. Energie und Politik im Anthropozän*. Berlin: Suhrkamp.

Däniken, Erich von. 1968. *Erinnerungen an die Zukunft*. Düsseldorf, Wien: Econ.

Däniken, Erich von. 1973. *Chariots of the Gods? Unsolved Mysteries of the Past*. News York: Bantam Books.

Daston, Lorraine, und Peter Galison. 2007. *Objektivität*. Frankfurt am Main: Suhrkamp.

Davies, Paul. 1999. Vorwort. In *Nachbarn im All. Auf der Suche nach Leben im Kosmos*, Hrsg. Seth Shostak, 9–14. München: Herbig.

Davies, Paul. 2007. ’Are Aliens Among Us?‘ *Scientific American* 297 (6): 62–69.

Davies, Paul & Robert Wagner. 2012. Searching for Alien Artifacts on the Moon. *Acta Astronautica* 89: 261–265.

Deardorff, James W. 1987. Examination of the Embargo Hypothesis as an Explanation for the Great Silence. *Journal of the British Interplanetary Society* 40: 373–379.

Denning, Kythrya. 2013. Impossible Predictions of the Unprecedented: Analogy, History and the Work of Prognostication. In *Astrobiology, History, and Society. Life Beyond Earth and the Impact of Discovery*, Hrsg. Douglas A. Vakoch, 301–312. Heidelberg: Springer.

de la Torre, Gabriel & Manuel A. Garcia. 2018. The Cosmic Gorilla Effect or the Problem of Undetected Non Terrestrial Intelligent Signals. *Acta Astronautica* 146: 83–91.

Dick, Steven J. 1982. *Plurality of Worlds. The Origins of Extraterrestrial Life Debate from Democritus to Kant*. Cambridge: Cambridge University Press.

Dick, Steven J. 1996. *The Biological Universe: The Twentieth-Century Extraterrestrial Life Debate and the Limits of Science*. Cambridge: Cambridge University Press.

Dick, Steven J. 2003. Cultural Evolution, the Postbiological Universe, and SETI. *International Journal of Astrobiology* 2 (1): 65–74.

Dick, Steven J. 2013. The Societal Impact of Extraterrestrial Life: The Relevance of History and the Social Sciences. In *Astrobiology, History, and Society. Life Beyond Earth and the Impact of Discovery*, Ed. Douglas A. Vakoch, 227–257. Heidelberg: Springer.

Dick, Steven J. 2014. Analogy and the Societal Implications of Astrobiology. *Astropolitics. The International Journal of Space Politics & Policy* 12: 210–230.

Dick, Steven J. (Ed.) 2015. *The Impact of Discovery Life beyond Earth.* Cambridge: University Press.

Dixon, Robert S. 2017. Statement Regarding the Claim that the „WOW!" Signal was Caused by Hydrogen Emission from an Unknown Comet or Comets. http://naapo.org/WOWCom etRebuttal.html.

Dodd, Adam. 2018. Strategic Ignorance and the Search for Extraterrestrial Intelligence: Critiquing the Discursive Segregation of UFOs from Scientific Inquiry. *Astropolitics. The International Journal of Space Politics and Policy* 16 (1): 75–95.

do Mar Castro Varala, María & Nikita Dhawan. (2. überarb. Aufl.) 2015. *Postkoloniale Theorie. Eine kritische Einführung.* Bielefeld: transcript.

Döring-Manteuffel, Sabine. 2008. *Das Okkulte. Eine Erfolgsgeschichte im Schatten der Aufklärung. Von Gutenberg bis zum World Wide Web.* München: Siedler.

Drake, Frank & Dava Sobel. 1992. *Is anyone out there? The Scientific Search for Extraterrestrial Intelligence.* New York: Delacorte Press.

Drake, Frank, & Dava Sobel. 1994. *Signale von anderen Welten. Die wissenschaftliche Suche nach außerirdischer Intelligenz.* München: Droemer.

Duhoux, Yves. 2000. How Not to Decipher the Phaistos Disc. A Review. *American Journal of Archaeology* 104: 597–600.

Ehman, Jerry R. 1998. The Big Ear Wow! Signal. What We Know and Don't Know About It After 20 Years. http://www.bigear.org/wow20th.htm#printout.

Elliott, John. 2009. A semantic 'engine' for universal translation. *Acta Astronautica* 68 (2011): 435-440.

Elliott, John. 2010. A post-detection decipherment strategy. *Acta Astronautica* 68 (2011): 441-444.

Elliott, John. 2011. Constructing the matrix. *Acta Astronautica* 78 (2012): 26-30.

Elliott, John. 2014. Beyond an Anthropomorphic Template. *Acta Astronautica* 116: 403–407.

Ellis, Erle C. 2018. *Anthropocence. A Very Short Introduction.* Oxford: Oxford University Press.

Engelbrecht, Martin. 2008a. SETI – Die wissenschaftliche Suche nach außerirdischer Intelligenz im Spannungsfeld divergierender Wirklichkeitskonzepte. In *Von Menschen und Außerirdischen. Transterrestrische Begegnungen im Spiegel der Kulturwissenschaften*, Hrsg. Michael Schetsche und Martin Engelbrecht, 205–226. Bielefeld: transcript.

Engelbrecht, Martin. 2008b. Von Aliens erzählen. In *Von Menschen und Außerirdischen. Transterrestrische Begegnungen im Spiegel der Kulturwissenschaft*, Hrsg. Michael Schetsche und Martin Engelbrecht, 13–29. Bielefeld: transcript.

Fetscher, Justus & Robert Stockhammer. 1997. Nachwort. In *Marsmenschen. Wie die Außerirdischen gesucht und erfunden wurden*, Hrsg. Justus Fetscher und Robert Sockhammer, 169–172. Leipzig: Reclam.

Fiedler, Peter. 2001. *Dissoziative Störungen und Konversion. Trauma und Traumabehandlung.* Beinheim: Beltz/PVU.

Fink, Alexander, und Andreas Siebe. 2006. *Handbuch Zukunftsmanagement. Werkzeuge der strategischen Planung und Früherkennung.* Frankfurt am Main: Campus.

Finney, Ben. 1990. The Impact of Contact. *Acta Astronautica* 21: 117–121

Finney, Ben and Jerry Bentley. 2014. A Tale of Two Analogues Learning at a Distance from the Ancient Greeks and Maya and the Problem of Deciphering Extraterrestrial Radio Transmissions. In *Archeology, Anthropology and Interstellar Communication,* Ed. Douglas Vakoch, 65–77. Washington: National Aeronautics and Space Administration.

Fischer, Joachim. 2009. *Philosophische Anthropologie. Eine Denkrichtung des 20. Jahrhunderts.* Freiburg im Breisgau: Karl Alber.

Fischer, Lars. 2017. Aus für Außerirdische. SPEKTRUM online: News 06.06.2017. https://www.spektrum.de/news/aus-fuer-ausserirdische/1462193.

Flechtheim, Ossip K. 1970. *Futurologie. Der Kampf um die Zukunft.* Köln: Verlag Wissenschaft und Politik.

Foster, G. V. 1972. Non-Human Artifacts in the Solar System. *Spaceflight* 14: 447–453.

Freitas Robert A. Jr. 1983. The Search for Extraterrestrial Artefacts (SETA). *Journal of the British Interplanetary Society* 36: 501–506.

Freitas, Robert A. Jr. 1985. The Search for Extraterrestrial Artifacts (SETA). *Acta Astronautica* 12: 1027–1034.

Freudenthal, Hans. 1960. *LINCOS. Design of a Language for Cosmic Intercourse.* Amsterdam: North-Holland Publishing.

Fuchs, Walter R. 1973. *Leben unter fernen Sonnen? Wissenschaft und Spekulation.* München: Knaur.

Garber, Stephen J. 1999. Searching for Good Science: The Cancellation of NASA's SETI Program. *Journal of the British Interplanetary Society* 52: 3–12.

Garber, Stephen, J. 2014. A Political History of NASA's SETI Program. In *Archeology, Anthropology and Interstellar Communication,* Ed. Douglas Vakoch, 23–48. Washington: National Aeronautics and Space Administration.

Gehlen, Arnold. 1986. *Der Mensch. Seine Natur und seine Stellung in der Welt.* Wiesbaden: Aula-Verlag.

Gerritzen, Daniel. 2016. *Erstkontakt. Warum wir uns auf Außerirdische vorbereiten müssen.* Stuttgart: Kosmos.

Giesen, Bernhard. 2010. *Zwischenlagen. Das Außerordentliche als Grund der sozialen Wirklichkeit.* Weilerswist: Vellbrück.

Grunwald, Martin. 2012. Das Sinnessystem Haut und sein Beitrag zur Körper-Grenzerfahrung. In *Körperkontakt. Interdisziplinäre Erkundungen,* Hrsg. Renate-Berenike Schmidt und Michael Schetsche, 29–54. Gießen: Psychosozial-Verlag.

Guthke, Karl S. 1983. *Der Mythos der Neuzeit. Das Thema der Mehrheit der Welten in der Literatur- und Geistesgeschichte von der kopernikanischen Wende bis zur Science Fiction.* Bern: Francke.

Günther, Ludwig. 1898. *Keplers Traum vom Mond.* Leipzig: Teubner.

Graf, Hans Georg. 2003. Was ist eigentlich Zukunftsforschung. *Sozialwissenschaft und Berufspraxis* 26 (4): 355-364.

Grazier, Kevin R. & Stephen Cass. 2015. *Hollyweird Science: From Quantum Quirks to the Multiverse.* New York: Springer.

Groh, Arnold. 1999. Globalisierung und kulturelle Information. In *Die Zukunft des Wissens. Workshop-Beiträge, XVIII. Deutscher Kongreß für Philosophie*, Hrsg. Jürgen Mittelstraß, 1076–1084. Konstanz: UVK.

Halbwachs, Maurice. 1967. *Das kollektive Gedächtnis*. Stuttgart: Enke.

Haqq-Misra, Jacob & Ravi Kumar Kopparapu. 2011. On the Likelihood of Non-Terrestrial Artifacts in the Solar System. *arXiv*: 1111.1212v1.

Harari, Yuval Noah. 2017. *Homo Deus. Eine Geschichte von Morgen*. München: C. H. Beck.

Harrison, Albert A. 1993. Thinking Intelligently about Extraterrestrial Intelligence: An Application of Living Systems Theory. *Behavioral Science* 38 (3): 189–217.

Harrison, Albert A. 1997. *After Contact. The Human Response to Extraterrestial Life*. New York, London: Plenum Trade.

Harrison, Albert A., und Alan C. Elms. 1990. Psychology and the search for extraterrestrial intelligence. *Behavioral Science* 35: 207–218.

Harrison, Albert A., und Joel T. Johnson. 2002. Leben mit Außerirdischen. In *S.E.T.I. Die Suche nach dem Außerirdischen*, Hrsg. Tobias Daniel Wabbel, 95–116. München: Beust.

Heidmann, Jean. 1994. *Bioastronomie. Über irdisches Leben und außerirdische Intelligenz*. Berlin: Springer.

Herrmann, Dieter B. 1988. *Rätsel um Sirius. Astronomische Bilder und Deutungen*. Berlin: Der Morgen.

Herzing, Denise L. 2014. Profiling Nonhuman Intelligence: An Exercise in Developing Unbiased Tools for Describing other „Types" of Intelligence on Earth. *Acta Astronautica* 94: 676–680.

Heuser, Marie-Luise. 2008. Transterrestrik in der Renaissance: Nikolaus von Kues, Giordano Bruno, Johannes Kepler. In *Von Menschen und Außerirdischen. Transterrestrische Begegnungen im Spiegel der Kulturwissenschaft*, Hrsg. Michael Schetsche und Martin Engelbrecht, 55–79. Bielefeld: transcript.

Hickman, John, und Koby Boatright. 2017. Stranger Danger: Extraterrestrial First Contact as Political Problem. *Space Review*. May 15., 2017.

Hickman, Leo. 2010. Stephen Hawking Takes a Hard Line on Aliens. *The Guardian*, April 26, 2010. https://www.theguardian.com/commentisfree/2010/apr/26/stephen-hawking-issues-warning-on-aliens.

Hilgartner, Stephen & Charles L. Bosk. 1988. The Rise and Fall of Social Problems: A Public Arenas Model. *American Journal of Sociology* 94: 53–78.

Hiroki, Kenzo. 2012. Strategies for Managing Low-probability, High-impact Events. Washington D. C.: World Bank. https://openknowledge.worldbank.org/handle/10986/16163.

Hitzler, Ronald & Michaela Pfadenhauer, Hrsg. 2005: *Gegenwärtige Zukünfte. Interpretative Beiträge zur sozialwissenschaftlichen Diagnose und Prognose*. Wiesbaden: VS Verlag.

Hoagland, Richard C. 1987. *The Monuments of Mars: A City on the Edge of Forever*. Berkeley: North Atlantic Books.

Hoerner, Sebastian von. 1967. Sind wir allein im Kosmos? *Neue Wissenschaft* 15 (1/2): 1–17.

Hoerner, Sebastian von. 2003. *Sind wir allein? SETI und das Leben im All*. München: C. H. Beck.

Hogrebe, Wolfram, Hrsg. 2005. *Mantik. Profile prognostischen Wissens in Wissenschaft und Kultur*. Würzburg: Königshausen und Neumann.

Höbel, Peter & Thorsten Hofmann. (2. völlig überarb. Aufl.) 2014. *Krisenkommunikation*. Konstanz: UVK.

Hölldobler, Bert & Edward Wilson. 2010. *Der Superorganismus. Der Erfolg von Ameisen, Bienen, Wespen und Termiten*. Berlin, Heidelberg: Springer.

Holzhauer, Hedda. 2015. *Kriminalistische Serendipity – Ermittlungserfolge im Spannungsfeld zwischen Berufserfahrung, Gefühlsarbeit und Zufallsentdeckungen*. (Dissertation, Universität Hamburg, Fachbereich Sozialwissenschaften).

Hövelmann, Gerd. 2008. Vernünftiges Reden und technische Rationalität. Erkenntnistheoretische Überlegungen zu Grundfragen der UFO-Forschung. In *Von Menschen und Außerirdischen. Transterrestrische Begegnungen im Spiegel der Kulturwissenschaft*, Hrsg. Michael Schetsche und Martin Engelbrecht, 183–204. Bielefeld: transcript.

Hövelmann, Gerd. 2009. Mutmaßungen über Außerirdische. *Zeitschrift für Anomalistik* 9: 168-199.

Hurst, Matthias. 2004. Stimmen aus dem All – Rufe aus der Seele. Kommunikation mit Außerirdischen in narrativen Spielfilmen. In *Der maximal Fremde. Begegnungen mit dem Nichtmenschlichen und die Grenzen des Verstehens*, Hrsg. Michael Schetsche, 95–112. Würzburg: Ergon.

Hurst, Matthias. 2008. Dialektik des Aliens. Darstellungen und Interpretationen von Außerirdischen in Film und Fernsehen. In *Von Menschen und Außerirdischen. Transterrestrische Begegnungen im Spiegel der Kulturwissenschaft*, Hrsg. Michael Schetsche und Martin Engelbrecht, 31–53. Bielefeld: transcript.

Huygens, Christiaan. 1703. *Cosmotheoros oder Eine phantastisch-realistische Betrachtung der Schönheit der Welt, der Sterne und Planeten. Geschrieben von Christiaan Huygens für seinen Bruder Constantijn, Geheimrat der königlichen Majestät von Großbritannien*. http://www.passagenproject.com/christiaan-huygens-cosmotheoros.html.

Hynek, J. Allen. 1972. *The UFO Experience. A Scientific Inquiry*. Chicago: Henry Regnery.

Hynek, J. Allen. 1979. *UFO. Begegnungen der ersten, zweiten und dritten Art*. München: Goldmann.

Janjic, Aleksandar. 2017. *Lebensraum Universum. Einführung in die Exoökologie*. Berlin: Springer.

Jastrow, Robert. 1997. What Are the Chances for Life? *Sky & Telescope* June 1997: 62–63.

Jones, Morris. 2013. Mainstream Media and Social Media Reactions to the Discovery of Extraterrestrial Life. In *Astrobiology, History and Society: Advances in Astrobiology and Biogeographics*, Ed. Douglas Vakoch, 313–328. Berlin: Springer.

Joshi, Manoj. 2003. Climate Model Studies of Synchronously Rotating Planets. *Astrobiology* 3 (2): 415–427.

Jüdt, Ingbert. 2013. Das UFO-Tabu ist öffentlich, nicht politisch. In *Diesseits der Denkverbote*, Hrsg. Michael Schetsche und Andreas Anton, 113–131. Hamburg: LIT.

Jungk, Robert. 1973. *Der Jahrtausendmensch. Bericht aus den Werkstätten der neuen Gesellschaft*. München: Bertelsmann.

Kaiser, Céline. 2004. „Fafagolik?" Fiktionen des Erstkontaktes in der ‚Marsliteratur' um 1900. In *Der maximal Fremde*, Hrsg. Michael Schetsche, 75–93. Würzburg: Ergon.

Kant, Immanuel. 1954. *Träume eines Geistersehers*. Berlin: Aufbau Verlag.

Kaplan, S. A. (Russian 1969) 1971. Exosociology - the Search for Signals from Extraterrestrial Civilisations. In *Extraterrestrial Civilizations. Problems of Interstellar Communications*, ed. S. A. Kaplan, 1–12. Jerusalem: Israel Program for Scientific Translations.

Kayser, Rainer. 2009. Spitzer und Hubble. Exoplanet mit organischen Molekülen. Astronews vom 21.10.2009.

Kennicutt, Robert C. & Neal J. Evans. 2012. Star Formation in the Milky Way and Nearby Galaxies. *Annual Review of Astronomy and Astrophysics* 50 (1): 531–608.

Kerner, Ina. 2012. *Postkoloniale Theorien zur Einführung*. Hamburg: Junius.

Knoblauch, Hubert & Bernt Schnettler. 2004. „Postsozialität", Alterität und Alienität. In *Der maximal Fremde. Begegnungen mit dem Nichtmenschlichen und die Grenzen des Verstehens*, Hrsg. Michael Schetsche, 23–42. Würzburg: Ergon.

Korhonen, Janne M. 2012. Mad with Aliens? Interstellar Deterrence and its Implications. *Acta Astronautica* 86: 201–210.

Koshland Daniel E. Jr. (2002). The Seven Pillars of Life. *Science* 295: 2215–2216.

Kosow, Hannah & Robert Gaßner. 2008. Methoden der Zukunfts- und Szenarioanalyse – Überblick, Bewertung und Auswahlkriterien (IZT-Werkstattbericht 103). Berlin: Institut für Zukunftsstudien und Technologiebewertung. https://www.izt.de/fileadmin/publikati onen/IZT_WB103.pdf.

Kramer, William M. & Charles W. Bahmer. (2. Aufl.) 1992. *Fire Officer's Guide to Disaster Control*. Tulsa, Oklahoma: Pennwell.

Krauss, Lawrence. 2002. Zahlenspiele mit Außerirdischen. In *Auf der Suche nach dem Außerirdischen*, Hrsg. Tobias Daniel Wabbel, 26–36. München: beustverlag.

Kreibich, Rolf. 2006. Zukunftsforschung (IZT-Arbeitsbericht 23/2006). http://www2.izt.de/ pdfs/IZT_AB_23.pdf.

Kues, Nikolaus von. 1985. *De docta ignoratia*. Book II.

Kuiper, Thomas B. H. & Mark Morris. 1977. Searching for Extraterrestrial Civilizations. *Science* 196: 616–621.

Kutschera, Ulrich. (4. Aufl.) 2015. *Evolutionsbiologie*. Stuttgart: UTB.

Leibundgut, Peter. 2011. *Ausserirdische und was Sie darüber wissen sollten*. Neckenmarkt (Österreich): Novum pro.

Lestel, Dominique. 2014. Ethology, Ethnology, and Communication with Extraterrestrial Intelligence. In *Archaeology, Anthropology, and Interstellar Communication*, Hrsg. Douglas A. Vakoch, 227–234. Washington: NASA.

Levin, Samuel R., Thomas W. Scott, Helen S. Cooper & Stuart A. West. 2017. Darwin's Aliens. *International Journal of Astrobiology*. https://doi.org/10.1017/S14735504170 00362. Zugegriffen: 15. Januar 2018.

Liu, Cixin. 2018. *Der dunkle Wald*. München: Heyne.

Locke, John. 1999 [1690]. *An Essay Concerning Human Understanding*. The Pennsylvania State University. http://www.philotextes.info/spip/IMG/pdf/essay_concerning_human_u nderstanding.pdf.

Lowric, Ian. 2013. Cultural Resources and Cognitive Frames: Keys to an Anthropological Approach to Prediction. In *Astrobiology, History, and Society. Life Beyond Earth and the Impact of Discovery*, Ed. Douglas A. Vakoch, 259–269. Heidelberg: Springer.http:// wlym.com/archive/pdf/cusa_learned02.pdf

Luhmann, Niklas. 1975. Die Weltgesellschaft. In *Soziologische Aufklärung*, Bd. 2., ders., 51–71. Wiesbaden: VS Verlag für Sozialwissenschaften.

Luhmann, Niklas. 1986. *Die Selbstbeschreibung der Gesellschaft und die Soziologie*. Gastvorlesung an der Universität Augsburg. Radiomitschnitt Bayerischer Rundfunk, 06.11.1986. https://www.youtube.com/watch?v=NIRTM3WZLqw.

Lynn, Steven Jay, Judith Pintar, Jane Stafford, Lisa Marmelstein & Timothy Lock. 1998. Rendering the Implausible Plausible: Narrative Construction, Suggestion, and Memory. In

Believed-in Imaginings: The Narrative Construction of Reality, Eds. Joseph de Rivera and Theodore R. Sarbin, 123–143. Washington: American Psychological Association.

Lynn, Steven Jay, Irving I. Kirsch. 1996. Alleged Alien Abductions: False Memory, Hypnosis, and Fantasy Proneness. *Psychological Inquiry* 7 (2): 151–155.

Marsiske, Hans-Arthur. 2005. *Heimat Weltall. Wohin soll die Raumfahrt führen?* Frankfurt am Main: Suhrkamp.

Marsiske, Hans-Arthur. 2007. Welche Sprache sprechen Außerirdische? Welt Online vom 09. Dezember 2007. https://www.welt.de/wissenschaft/article1439767/Welche-Sprache-sprechen-Ausserirdische.html.

Martinez, Claudio L. Flores. 2014. SETI in the Light of Cosmic Convergent Evolution. *Acta Astronautica* 104: 341–349

Maul, Stefan. 2013. *Die Wahrsagekunst im alten Orient*. München: Beck.

Mayer, Gerhard. 2003. Über Grenzen schreiben. Presseberichterstattung zu Themen aus dem Bereich der Anomalistik und der Grenzgebiete der Psychologie in den Printmedien SPIEGEL, BILD und BILD AM SONNTAG. *Zeitschrift für Anomalistik* 3: 8–46.

Mayor, Michel & Didier Queloz. 1995. A Jupiter-Mass Companion to a Solar-Type Star. *Nature* 378: 355–359.

Mayr, Ernst. 1995. Space Topics: Search for Extraterrestrial Intelligence. https://web.arc hive.org/web/20081115225902/http://www.planetary.org/explore/topics/search_for_life/seti/mayr.html.

McConnell, Brian. 2001. *Beyond Contact. A Guide to SETI and Communicating with Alien Civilisation*. Sebastopol, CA: O'Reilly.

McLeod, Caroline C., Barbara Corbisier & John E. Mack. 1996. A More Parsimonious Explanation for UFO Abduction. *Psychological Inquiry* 7 (2): 156–168.

Meadows, Dennis L. 1972. *Die Grenzen des Wachstums. Bericht des Club of Rome zur Lage der Menschheit*. Stuttgart: Deutsche Verlags-Anstalt.

Meadows, Dennis et al. 1972. The *Limits to Growth. A Report for the Club of Rome's Project on the Predicament of Mankind*. New York: Universe Books

Meierhenrich, Uwe J., Guillermo M. Munoz Caro, Jan Hendrik Bredehöft, Elmar K. Jessberger & Wolfram H.-P. Thiemann. 2004. Identification of diamino acids in the Murchison meteorite. *Proceedings of the National Academy of Sciences of the United States of America* 101 (25): 9182–9186.

Mejer, Jan H. 1983. Towards an Exo-Sociology: Constructs of the Alien. *Free Inquiry in Creative Sociology* 11 (2): 171–174.

Michaud, Michael A. G. 1972. Interstellar Negotiation. *Foreign Service Journal* Dec. 1972: 10–20.

Michaud, Michael A. G. 1999. A Unique Moment in Human History. In *Are we Alone in the Cosmos? The Search for Alien Contact in the New Millennium*, Ed. Byron Preiss und Ben Bova, 265–284. New York: ibooks.

Michaud, Michael A. G. 2007a. *Contact with Alien Civilizations. Our Hopes and Fears about Encountering Extraterrestrials*. New York: Springer.

Michaud, Michael A. G. 2007b. Ten Decisions that Could Shake the World. http://avsport.org/IAA/decision.pdf.

Michaud, Michael A. G. 2015. Searching for Extraterrestrial Intelligence: Preparing for an Expected Paradigm Break. In *The Impact of Discovery Life beyond Earth*, Ed. Steven J. Dick, 286–298. Cambridge: University Press.

Moore, Ben. 2014. *Da draußen. Leben auf unserem Planeten und anderswo.* Zürich: Kein & Aber.

Moore, Matthew. 2008. Messages from Earth Sent to Distant Planet by Bebo. *The Telegraph,* October 9, 2008. https://www.telegraph.co.uk/news/newstopics/howaboutthat/3166709/ Messages-from-Earth-sent-to-distant-planet-by-Bebo.html.

Morris, Simon Conway. 2003. The Navigation of Biological Hyperspace. *International Journal of Astrobiology* 2 (2): 149–152.

Müller, Klaus E. 2003. Tod und Auferstehung. Heilserwartungsbewegungen in traditionellen Gesellschaften. In *Historische Wendeprozesse. Ideen, die Geschichte machten,* Hrsg. Klaus E. Müller, 256–287. Freiburg: Herder.

Müller, Klaus E. 2004. Einfälle aus einer anderen Welt. In *Der maximal Fremde. Begegnungen mit dem Nichtmenschlichen und die Grenzen des Verstehens,* Hrsg. Michael Schetsche, 191–204. Würzburg: Ergon.

Münkler, Herfried & Bernd Ladwig. 1997. Dimensionen der Fremdheit. In *Furcht und Faszination. Facetten der Fremdheit,* Hrsg. Herfried Münkler, 11–43. Berlin: Akademie Verlag.

Nagel, Thomas. 2014. *Mind and Cosmos. Why the Materialist Neo-Darwinian Conception of Nature is Almost Certainly False.* Oxford: University Press.

Neal, Mark. 2014. Preparing for Extraterrestrial Contact. *Risk Management* 16 (2): 63–87.

Newman, Leonard S. & Roy F. Baumeister. 1996. Toward an Explanation of UFO Abduction Phenomenon: Hypnotic Elaboration, Extraterrestrial Sadomasochism, and Spurious Memories. *Psychological Inquiry* 7 (2): 99–126.

Neyer, Franz J. & Frank M. Spinath, Hrsg. 2008. *Anlage und Umwelt. Neue Perspektiven der Verhaltensgenetik und Evolutionspsychologie.* Stuttgart: Lucius & Lucius.

Nolting, Tobias & Ansgar Thießen. (Hrsg.) 2008. *Krisenmanagement in der Mediengesellschaft. Potenziale und Perspektiven in der Krisenkommunikation.* Wiesbaden: VS Verlag für Sozialwissenschaften.

Oeser, Erhard. 2009. *Die Suche nach der zweiten Erde. Illusion und Wirklichkeit der Weltraumforschung.* Darmstadt: WGB.

Ollongren, Alexander. 2010. On the Signature of LINCOS. *Acta Astronautica* 67: 1440–1442.

Orne, Martin M., Wayne G. Whitehouse, Emily Carota Orne & David F. Dinges. 1996., Memories of Anomalous and Traumatic Autobiographical Experiences: Validation and Consolidation of Fantasy through Hypnosis. *Psychological Inquiry* 7 (2): 168–172.

Paley, John. 1997. Satanist Abuse and Alien abduction: A Comparative Analysis Theorizing Temporal Lobe Activity as a Possible Connection between Anomalous Memories. *The British Journal of Social Work* 27: 43–70.

Panovkin, Boris Nikolaevich. 1976. The Objectivity of Knowledge and the Problem of the Exchange of Coherent Information with Extraterrestrial Civilizations. *Philosophical Problems of 20th Century Astronomy* (Moscow: Russian Academy of Sciences): 240–265.

Paris, Antonio. 2017. Hydrogen Line Observations of Cometary Spectra at 1420 MHZ. *Journal of the Washington Academy of Sciences 103 (2).* http://planetary-science.org/wp-content/uploads/2017/06/Paris_WAS_103_02.pdf.

Peters, Ted. 2013. Would the Discovery of ETI Provoke a Religious Crisis? In *Astrobiology, History and Society: Advances in Astrobiology and Biogeographics,* Ed. Douglas Vakoch, 341–355. Berlin: Springer.

Petigura, Erik A., Andrew W. Howard and Geoffrey W. Marcy. 2013. Prevalence of Earth-Size Planets Orbiting Sun-Like Stars. *Proceedings of the National Academy of Sciences of the United States of America* 110: 19273–19278.

Pirschl, Julia and Michael Schetsche. 2013. Aus Fehlern lernen. Anthropozentrische Vorannahmen im SETI-Paradigma – Folgerungen für die UFO-Forschung. In *Diesseits der Denkverbote. Bausteine für eine reflexive UFO-Forschung*, Hrsg. Michael Schetsche und Andreas Anton, 29–48. Berlin: Lit-Verlag.

Plutarch. 1968. *Das Mondgesicht (De facie in orbe lunae)*. Eingeleitet, übersetzt und erläutert von Herwig Görgemanns. Zürich: Artemis Verlag.

Pooley, Jefferson D. 2013. Checking Up on The Invasion from Mars: Hadley Cantril, Paul Felix Lazarsfeld, and the Making of a Misremembered Classic. *International Journal of Communication* 7: 1920–1948.

Porter, Jennifer E. 1996. Spiritualists, Aliens and UFOs: Extraterrestrials as Spirit Guides. *Journal of Contemporary Religion* 11: 337–353.

Race, Margaret. 2015. Preparing for the Discovery of Extraterrestrial Life: Are we Ready? Considering Potential Risks, Impacts, and Plans. In *The Impact of Discovery Life beyond Earth*, Ed. Steven J. Dick, 263–285. Cambridge: University Press.

Rausch, Renate. 1992. Der Kulturschock der Indios. In *1492 und die Folgen: Beiträge zur interdisziplinären Ringvorlesung an der Philipps-Universität Marburg*, Hrsg. Hans-Jürgen Prien, 18–32. Münster, Hamburg: LIT.

Raybeck, Douglas. 2014a. Predator-Prey Models and Contact Considerations. In *Extraterrestrial Altruism. Evolution and Ethics in the Cosmos*, Ed. Douglas A. Vakoch, 49–63. Berlin, Heidelberg: Springer.

Raybeck, Douglas. 2014b. Contact Considerations a Cross-Cultural Perspective. In *Archaeology, Anthropology, and Interstellar Communication*, Ed. Douglas A. Vakoch, 142–158. Washington: NASA.

Richter, Jonas. 2015. Paläo-SETI. In *An den Grenzen der Erkenntnis. Handbuch der wissenschaftlichen Anomalistik*, Hrsg. Gerhard Mayer, Michael Schetsche, Ina Schmied-Knittel, und Dieter Vaitl, 346–347. Stuttgart: Schattauer.

Richter, Jonas. 2017. *Götter-Astronauten. Erich von Däniken und die Paläo-SETI-Mythologie*. Hamburg: Lit.

Robitaille, Thomas P. and Barbara A. Whitney. 2010. The Present-Day Star Formation Rate of the Milky Way Determined from Spitzer-Detected Young Stellar Objects. *The Astrophysical Journal Letters* 710 (1): L11–L15.

Romesberg, Daniel Ray. 1992. *The Scientific Search for Extraterrestrial Intelligence: A Sociological Analysis*. Ann Arbor: UMI Dissertation Services.

Rummel, John D. 2001. Planetary Exploration in the Time of Astrobiology: Protecting against Biological Contamination. *PNAS* 98 (5): 2128–2131.

Sagan, Carl, und Iossif Samuilowitsch Schklowski. 1966. *Intelligent Life in the Universe*. San Francisco: Holden-Day.

Sample, Ian. 2018. Nasa's Golden Record May Baffle Alien Life, Say Researchers. *The Guardian*, May 26., 2018. https://www.theguardian.com/science/2018/may/26/nasas-gol den-record-may-baffle-alien-life-say-researchers.

Scalo, John, Lisa Kaltenegger, Antígona Segura, Malcolm Fridlund, Ignasi Ribas, Yu. N. Kulikov, John L. Grenfell, Heike Rauer, Petra Odert, Martin Leitzinger, Franck Selsis, Maxim L. Khodachenko, Carlos Eiroa, Jim Kasting and Helmut Lammer. 2007. M Stars as Targets for Terrestrial Exoplanet Searches and Biosignature Detection. *Astrobiology* 7 (1), 85–166.

Schacter, Daniel L. 2001. *Wir sind Erinnerung. Gedächtnis und Persönlichkeit.* Reinbek bei Hamburg: Rowohlt.

Scheler, Max. 1924. *Versuche zu einer Soziologie des Wissens.* Leipzig, Berlin: Duncker & Humblot.

Scheler, Max. (Orig. 1928) 2016. *Die Stellung des Menschen im Kosmos.* Berlin: Contumax.

Schetsche, Michael. 1997. „Entführungen durch Außerirdische" – ein ganz irdisches Deutungsmuster. *Soziale Wirklichkeit* 1: 259–277.

Schetsche, Michael. 2000. *Wissenssoziologie sozialer Probleme. Grundlegung einer relativistischen Problemtheorie.* Wiesbaden: Westdeutscher Verlag.

Schetsche, Michael. 2003. Soziale Folgen der Entdeckung einer außerirdischen Zivilisation (dreiteilig*). Nachrichten der Olbers-Gesellschaft,* Teil 1, Heft 200 (Januar 2003): 33–37; Teil 2, Heft 202 (Juli 2003): 26–30; Teil 3, Heft 203: 7–11.

Schetsche, Michael. 2004. Der maximal Fremde - eine Hinführung. In *Der maximal Fremde. Begegnungen mit dem Nichtmenschlichen und die Grenzen des Verstehens,* Hrsg. Michael Schetsche, 13–21. Würzburg: Ergon.

Schetsche, Michael. 2005a. Rücksturz zur Erde? Zur Legitimierung und Legitimität der bemannten Raumfahrt. In *Rückkehr ins All* (Ausstellungskatalog, Kunsthalle Hamburg), 24–27. Ostfildern: Hatje Cantz.

Schetsche Michael. 2005a: Zur Prognostizierbarkeit der Folgen außergewöhnlicher Ereignisse. In: Gegenwärtige *Zukünfte. Interpretative Beiträge zur sozialwissenschaftlichen Diagnose und Prognose,* Hrsg. Ronald Hitzler und Michaela Pfadenhauer, 55–71. Wiesbaden: Springer VS.

Schetsche, Michael. 2005b. Rücksturz zur Erde? Zur Legitimierung und Legitimität der bemannten Raumfahrt. In *Rückkehr ins All* (Ausstellungskatalog, Kunsthalle Hamburg), 24–27. Ostfildern: Hatje Cantz.

Schetsche, Michael. 2008a. Auge in Auge mit dem maximal Fremden? Kontaktszenarien aus soziologischer Sicht. In *Von Menschen und Außerirdischen. Transterrestrische Begegnungen im Spiegel der Kulturwissenschaft,* Hrsg. Michael Schetsche und Martin Engelbrecht, 227–253. Bielefeld: transcript.

Schetsche, Michael. 2008b. Das Geheimnis als Wissensform. Soziologische Anmerkungen. *Journal for Intelligence, Propaganda and Security Studies* 2 (1): 33–50.

Schetsche, Michael. 2008c. Entführt! Von irdischen Opfern und außerirdischen Tätern. In *Von Menschen und Außerirdischen. Transterrestrische Begegnungen im Spiegel der Kulturwissenschaft,* Hrsg. Michael Schetsche und Martin Engelbrecht, 157–182. Bielefeld: transcript.

Schetsche, Michael. 2008d. Der maximal Fremde – eine Hinführung. In *Der maximal Fremde. Begegnungen mit dem Nichtmenschlichen und die Grenzen des Verstehens,* Hrsg. Michael Schetsche, 13–21. Würzburg: Ergon.

Schetsche, Michael. 2012. Theorie der Kryptodoxie. Erkundungen in den Schattenzonen der Wissensordnung. *Soziale Welt* 63 (1): 5–25.

Schetsche, Michael. 2013. Unerwünschte Wirklichkeit. Individuelle Erfahrung und gesellschaftlicher Umgang mit dem Para-Normalen heute. *Zeitschrift für Historische Anthropologie* 21: 387–402.

Schetsche, Michael. 2015. Anomalien im medialen Diskurs. In *An den Grenzen der Erkenntnis. Handbuch der wissenschaftlichen Anomalistik*, Hrsg. Gerhard Mayer, Michael Schetsche, Ina Schmied-Knittel und Dieter Vaitl, 63–73. Stuttgart: Schattauer.

Schetsche, Michael & Andreas Anton. 2013. Einleitung: Diesseits der Denkverbote. In *Diesseits der Denkverbote. Bausteine für eine reflexive UFO-Forschung*, Hrsg. Michael Schetsche und Andreas Anton, 7–27. Hamburg: Lit.

Schetsche, Michael & Ina Schmied-Knittel. 2018a. Zur Einleitung: Heterodoxien in der *Moderne*. In *Heterodoxie. Konzepte, Traditionen, Figuren der Abweichung*, Hrsg. Michael Schetsche und Ina Schmied-Knittel, S. 9–33. Köln: Herbert von Halem.

Schetsche, Michael & Ina Schmied-Knittel. (Hrsg.) 2018b. *Heterodoxie. Konzepte, Traditionen, Figuren der Abweichung*. Köln: Herbert von Halem.

Schmied-Knittel, Ina & Edgar Wunder. 2008. UFO-Sichtungen. Ein Versuch der Erklärung äußerst menschlicher Erfahrungen. In *Von Menschen und Außerirdischen. Transterrestrische Begegnungen im Spiegel der Kulturwissenschaft*, Hrsg. Michael Schetsche und Martin Engelbrecht, 133–155. Bielefeld: transcript.

Schmied-Knittel, Ina & Michael Schetsche. 2003. Psi-Report Deutschland. Eine repräsentative Bevölkerungsumfrage zu außergewöhnlichen Erfahrungen. In *Alltägliche Wunder. Erfahrungen mit dem Übersinnlichen – wissenschaftliche Befunde*, Hrsg. Eberhard Bauer und Michael Schetsche, 13–38. Würzburg: Ergon.

Schetsche, Michael, René Gründer, Gerhard Mayer & Ina Schmied-Knittel. 2009. Der maximal Fremde. Überlegungen zu einer transhumanen Handlungstheorie. *Berliner Journal für Soziologie* 19 (3): 469–491.

Schmitt, Stefan. 2017. Das Wir da draußen. Wer nach Außerirdischen sucht, der findet – den Menschen. Im Kino, wo ein neuer ‚Alien-Film' anläuft. Aber auch in der Astronomie. *Die Zeit* Nr. 20 vom 11. Mai 2017: 37.

Schmitz, Michael. 1997. *Kommunikation und Außerirdisches. Überlegungen zur wissenschaftlichen Frage nach Verständigung mit außerirdischer Intelligenz*. (Unveröffentlichte Magisterarbeit, Universität-Gesamthochschule Essen).

Schnabel, Jim. 1994. Chronicles of Aliens Abduction and some other Traumas as Self-Victimization Syndrom. *Dissociation: Progress in the Dissociative Disorders* 7 (1): 51–62.

Scholz, Mathias. 2014. *Planetologie extrasolarer Planeten*. Heidelberg: Springer Spektrum.

Schrogl, Kai-Uwe. 2008. Weltraumpolitik, Weltraumrecht und Außerirdische(s). In *Von Menschen und Außerirdischen. Transterrestrische Begegnungen im Spiegel der Kulturwissenschaft*, Hrsg. Michael Schetsche und Martin Engelbrecht, 255–266. Bielefeld: transcript.

Schulze-Makuch, Dirk. 2017. Forty Years Later, SETI's Famous Wow! Signal May Have an Explanation. But the controversy continues (06.08.2017). https://www.airspacemag.com/daily-planet/forty-years-later-setis-famous-wow-signal-may-have-explanation-180963628/.

Schulze-Makuch, Dirk. 2018. How to Communicate with Aliens. Some Interesting Ideas Bounced Around at a Recent Workshop. Air & Space Smithsonian. https://www.airspa cemag.com/daily-planet/how-communicate-aliens-180969211/.

Schulze-Makuch, Dirk and Luis N. Irwin. 2004. *Life in the Universe. Expectations and Constraints.* Heidelberg: Springer.

Schulze-Makuch, Dirk and William Bains. 2017. *The Cosmic Zoo. Complex Life on Many Worlds.* Cham: Springer Nature.

Selsis, Franck, James F. Kasting, Benjamin Levrard, Jimmy Paillet, Ignasi Ribas and Xavier Delfosse. 2008. Habitable Planets Around the Star Gl 581? *Astronomy & Astrophysics.* https://arxiv.org/pdf/0710.5294.pdf. Zugegriffen: 14. März 2018).

Sheridan, Mark A. 2009. *SETI's Scope: How the Search for ExtraTerrestrial Intelligence Became Disconnected from New Ideas about Extraterrestrials.* Ann Arbor, MI: ProQuest.

Shermer, Michael. 2002. Why ET Hasn't Called. Scientific American. https://michaelsh ermer.com/2002/08/why-et-hasnt-called/.

Shostak, Seth. 1998. *Sharing the Universe. Perspectives on Extraterrestrial Life.* Berkeley: Berkeley Hills Books.

Shostak, Seth. 1999. *Nachbarn im All. Auf der Suche nach Leben im Kosmos.* München: Herbig.

Shostak, Seth. 2006. The Future of SETI. Sky and Telescope online (19.06.2006). http://www.skyandtelescope.com/astronomy-news/the-future-of-seti/3/?c=y.

Shostak, Seth. 2015. Searching for Clever Life. *Astrobiology* 15 (11): 948–950.

Showalter, Elaine. 1997. *Hystorien. Hysterische Epidemien im Zeitalter der Medien.* Berlin: Berlin Verlag.

Shuch, H. Paul. 2011. *Searching for Extraterrestrial Intelligence - SETI Past, Present, and Future.* Berlin: Springer.

Simmel, Georg. (4. Aufl.) 1958. *Soziologie: Untersuchungen über die Formen der Vergesellschaftung.* Berlin: Duncker & Humblot.

Simons, Daniel J. & Christopher F. Chabris. 1999. Gorillas in our Midst: Sustained Inattentional Blindness for Dynamic Events. *Perception* 28: 1059–1074.

Smart, John M. 2012. The Transcension Hypothesis: Sufficiently Advanced Civilizations in Variably Leave our Universe and Implications for METI and SETI. *Acta Astronautica* 78: 55–68.

Smith, John Maynard & Eörs Szathmáry. 1995. *The Major Transitions in Evolution.* Oxford, New York: Freeman and Company.

Spanos, Nicholas P., Cheryl A. Burgess & Melissa Faith. 1994. Past-life Identity, UFO Abductions, and Satanic Ritual Abuse: The Social Construction of Memories. *The International Journal of Clinical and Experimental Hypnosis* XLII (4): 433–446

Spanos, Nicholas P., Patricia A. Cross, Kirby Dickson & Susan C. DuBreuil. 1993. Close Encounters: An Examination of UFO Experiences. *Journal of Abnormal Psychology* 102: 624–632.

Spreen, Dierk, und Joachim Fischer. 2014. *Soziologie der Weltraumfahrt.* Bielefeld: transcript.

Spry, Andy J. 2009. Contamination Control and Planetary Protection. In *Drilling in Extreme Environments – Penetration and Sampling on Earth and other Planets*, Ed. Yoseph Bar-Cohen & Kris Zacny, 707–739. Weinheim: Wiley-VCH.

Stagl, Justin. 1981. Die Beschreibung des Fremden in der Wissenschaft. In *Der Wissenschaftler und das Irrationale*, zweiter Band, Hrsg. Hans Peter Duerr, 273–295. Frankfurt am Main: Syndikat.

Stagl, Justin. 1997. Grade der Fremdheit. In *Furcht und Faszination – Facetten der Fremdheit*, Hrsg. Herfried Münkler, 85-114. Berlin: Akademie Verlag.

Steinmüller, Angela & Heinz Steinmüller. (2. Aufl.) 2004. *Wild Cards. Wenn das Unwahrscheinliche eintritt*. Hamburg: Murmann.

Stenger, Horst. 1998. Soziale und kulturelle Fremdheit. Zur Differenzierung von Fremdheitserfahrungen am Beispiel ostdeutscher Wissenschaftler. *Zeitschrift für Soziologie* 27 (1): 18–38.

Stevenson, David S. & Sean Large. 2017. Evolutionary Exobiology: Towards the Qualitative Assessment of Biological Potential on Exoplanets. *International Journal of Astrobiology*. https://doi.org/10.1017/S1473550417000349. Zugegriffen: 18. Januar 2018

Stirn, Alexander. 2016. Flotte von Mini-Raumschiffen soll zu Alpha Centauri fliegen. Süddeutsche Zeitung, 13.04.2016. http://www.sueddeutsche.de/wissen/breakthrough-sta rshot-flotte-von-mini-raumschiffen-soll-zu alpha-centauri-fliegen-1.2947852.

Storch, Volker, Ulrich Welsch & Michael Wink. (3. Aufl.) 2013. *Evolutionsbiologie*. Heidelberg: Springer.

Streeck-Fischer, Annette, Ulrich Sachsse & Ibrahim Özkan. 2001. Perspektiven in der Traumaforschung. In *Körper, Seele, Trauma*, Hrsg. Annette Streeck-Fischer, Ulrich Sachsse und Ibrahim Özkan, 12–22. Göttingen: Vandenhoeck & Ruprecht.

Strugazki, Arkade & Boris Strugazki. (russ. Orig. 1971) 1975. *Picknick am Wegesrand. Utopische Erzählung*. Berlin: Verlag Das neue Berlin.

Stuckrad, Kocku von. 2007. *Geschichte der Astrologie. Von den Anfängen bis zur Gegenwart*. München: Beck.

Sukhotin, Boris Viktorovich. 1971. Methods of Message Decoding. In *Extraterrestrial Civilizations. Problems of Interstellar Communication*, Ed. S. A. Kaplan,133–212. Jerusalem: Keter Press.

Teltsch, Kathleen. 1977. U. N. Sending Messages Aboard Voyager Craft for Beings in Space. *New York Times*, June 3, 1977. https://www.nytimes.com/1977/06/03/archives/un-sen ding-messages-aboard-voyager-craft-for-beings-in-space.html.

Toepfer, Georg. 2011. Leben. In *Historisches Wörterbuch der Biologie. Geschichte und Theorie der biologischen Grundbegriffe*, Bd. 2. Stuttgart: Metzler: 420–483.

Traphagen, John W. 2014. Culture and Communication with Extraterrestrial Intelligence. In *Archaeology, Anthropology, and Interstellar Communication*, Hrsg. Douglas A. Vakoch, 159–172. Washington: NASA.

Uerz, Gereon. 2006. *ÜberMorgen. Zukunftsvorstellungen als Elemente der gesellschaftlichen Konstruktion der Wirklichkeit*. Paderborn, München: Fink.

Ulbrich Zürni, Susanne. 2004. *Möglichkeiten und Grenzen der Szenarioanalyse*. Stuttgart, Berlin: WiKu.

Urban, Tim. 2014. The Fermi Paradox. http://waitbutwhy.com/2014/05/fermi-paradox.html.

Urban, Tim. 2015. The AI Revolution: Our Immortality or Extinction. https://waitbutwhy. com/2015/01/artificial-intelligence-revolution-1.html and https://waitbutwhy.com/2015/ 01/artificial-intelligence-revolution-2.html.

Vacarr, Barbara Adina. 1993. *The Divine Container. A Transpersonal Approach in the Treatment of Repressed Abduction Trauma.* Ph.D., The Union Institute (Cincinnati/Ohio). Demand Copy: University Microfilms International.

Vakoch, Douglas A. 2011. Asymmetry in Active SETI: A Case for Transmissions from Earth. *Acta Astronautica* 68: 476–488.

Vakoch, Douglas A. 2014a. Archeology, Anthropology and Interstellar Communication. Washington: National Aeronautics and Space Administration. https://www.nasa.gov/sites/default/files/files/Archaeology_Anthropology_and_Interstellar_Communication_TAGGED.pdf.

Vakoch, Douglas A. 2014b. The Evolution of Extraterrestrials. The Evolutionary Synthesis and Estimates of the Prevalence of Intelligence beyond Earth. In *Archaeology, Anthropology, and Interstellar Communication,* Ed. Douglas A. Vakoch, 189–202. Washington: NASA.

Vossenkuhl, Wilhelm. 1990. Jenseits des Vertrauten und Fremden. In *Einheit und Vielfalt.* XIV. Dt. Kongress für Philosophie Giessen, 21.-26. September 1987, Hrsg. Odo Marquard, 101–113. Hamburg: Meiner.

Wabbel, Tobias Daniel. 2002. Der Geist des Radios. In *S.E.T.I. Die Suche nach dem Außerirdischen,* Hrsg. Tobias Daniel Wabbel, 67–79. München: beustverlag.

Waldenfels, Bernhard. 1997. *Topographie des Fremden. Studien zur Phänomenologie des Fremden,* Band 1. Frankfurt am Main: Suhrkamp.

Walter, Ulrich. 2001. *Außerirdische und Astronauten. Zivilisationen im All.* Heidelberg: Spektrum Akademischer Verlag.

Wandel, Amri. 2014. On the Abundance of Extraterrestrial Life After the Kepler Mission. *International Journal of Astrobiology* 14 (3): 511–516.

Ward, Peter. 2009. Gaias böse Schwester. *Spektrum der Wissenschaft* 11: 84–88.

Wason, Paul K. 2014. Inferring Intelligence. Prehistoric and Extraterrestrial. In *Archaeology, Anthropology, and Interstellar Communication,* Ed. Douglas A. Vakoch, 112–128. Washington: NASA.

Wells, Herbert George. 1999 [1898]. *The War of the Worlds.* https://gutenberg.org/files/36/36-h/36-h.htm.

Wendt, Alexander & Raymond Duvall. 2008. Sovereignty and the UFO. *Political Theory* 36 (4): 607–633.

Wendt, Alexander & Raymond Duvall. 2012. Militanter Agnostizismus und das UFO-Tabu. In *Generäle, Piloten und Regierungsvertreter brechen ihr Schweigen,* Hrsg. Leslie Kean, 281–294. Rottenburg: Kopp.

Werthimer, Dan, David NG, Stuart Bowyer, und Charles Donnelly. 1995. The Berkeley SETI Program: SERENDIP III and IV Instrumentation. In *Progress in the Search for Extraterrestrial Life,* Ed. Seth Shostak, Astronomical Society of the Pacific Conference Series 74: 293–302.

Weyer, Johannes. 1997. Technikfolgenabschätzung in der Raumfahrt. In *Technikfolgenabschätzung als politische Aufgabe,* Hrsg. Raban Graf von Westphalen, 465–483. München: Oldenburg.

Whitmore, John. 1993. Religious Dimensions of the UFO Abductee Experience. *Syzygy: Journal of Alternative Religion and Culture* 2 (3–4): 313–326.

Wille, Holger. 2005. *Kant über Außerirdische. Zur Figur des Alien im vorkritischen und kritischen Werk.* Münster: Monsenstein & Vannerdat.

Wirth, Sven et al., Hrsg. 2016. *Das Handeln der Tiere. Tierische Agency im Fokus der Human-Animal-Studies*. Bielefeld: transcript.
Woolgar, Steve & Dorothee Pawluch. 1985. Ontological Gerrymandering: The Anatomy of Social Problems Explanations. *Social Problems* 32: 214–227.
Zackrisson, Erik, Andreas J. Korn, Ansgar Wehrhahn & Johannes Reiter. 2018. SETI with Gaia: The Observational Signatures of Nearly Complete Dyson Spheres. https://arxiv.org/abs/1804.08351.
Zaitsev, Aleksandr L. 2006. Messaging to Extra-Terrestrial Intelligence. arXiv:physics/061 0031. https://arxiv.org/ftp/physics/papers/0610/0610031.pdf.
Zaitsev, Aleksandr L. 2008. The First Musical Interstellar Radio Message. *Journal of Communications Technology and Electronics* 53 (9): 1107–1113.
Zaun, Harald. 2006. Bewohnte Welten um Rote Zwergsterne? Telepolis (Online-Magazin). https://www.heise.de/tp/features/Bewohnte-Welten-um-Rote-Zwergsterne-3404750.html.
Zaun, Harald. (Hrsg.) 2010. *Kosmologie – Intelligenzen im All*. Hannover: Heise.
Zaun, Harald. 2010. *SETI – Die wissenschaftliche Suche nach außerirdischen Zivilisationen. Chancen, Perspektiven, Risiken*. Hannover: Heise.
Zaun, Harald. 2015. „Dieses neue SETI-Programm stellt alles Bisherige in den Schatten!" Telepolis am 21. Juli 2015. https://www.heise.de/tp/features/Dieses-neue-SETI-Programm-stellt-alles-Bisherige-in-den-Schatten-3374394.html?seite=all.
Zaun, Harald. 2017. Historisches SETI-Signal ohne Kosmogram. Telepolis am 15. August 2017. https://www.heise.de/tp/features/Historisches-SETI-Signal-ohne-Kosmogramm-3801610.html.

First Contact Science Fiction Movies

The Day the Earth Stood Still (Robert Wise, USA 1951)
The War of the Worlds (Byron Haskin, USA 1953)
Invasion of the Body Snatchers (Don Siegel, USA 1956)
The Blob (Irvin S. Yeaworth junior, USA 1958)
The Village of the Damned (Wolf Rilla, UK 1960)
2001: A Space Odyssey (Stanley Kubrick, UK/USA 1968)
Solaris (Andrei Arsenyevich Tarkovsky, USSR 1972)
Close Encounters of the Third Kind (Steven Spielberg, USA 1977)
Stalker (Andrei Arsenyevich Tarkovsky, USSR 1979)
The Abyss (James Cameron, USA 1989)
Contact (Robert Lee Zemeckis, USA 1997)
Sphere (Barry Levinson, USA 1998)
K-PAX (Iain Softley, UK/Germany/USA 2001)
District 9 (Neill Blomkamp, USA/New Zealand/Canada/South Africa 2009)
Oblivion (Joseph Kosinski, USA 2013)
Arrival (Denis Villeneuve, USA 2016)
Life (Daniél Espinosa, USA 2017)
Annihilation (Alex Garland, USA/UK 2018)
A Quiet Place (John Krasinski, USA 2018)